T0295581

Analyzing Electoral Promises with Game Theory

Electoral promises help to win votes and political candidates, or parties should strategically choose what they can deliver to win an election. Past game-theoretical studies tend to ignore electoral promises and this book sheds illuminating light on the functions and effects of electoral promises on policies or electoral outcomes through game theory models. This book provides a basic framework for game-theoretical analysis of electoral promises.

The book also includes cases to illustrate real life applications of these theories.

Yasushi Asako is Associate Professor at the Faculty of Political Science and Economics, Waseda University.

Routledge Focus on Economics and Finance

The fields of economics are constantly expanding and evolving. This growth presents challenges for readers trying to keep up with the latest important insights. Routledge Focus on Economics and Finance presents short books on the latest big topics, linking in with the most cutting-edge economics research.

Individually, each title in the series provides coverage of a key academic topic, whilst collectively the series forms a comprehensive collection across the whole spectrum of economics.

Knowledge Infrastructure and Higher Education in India
Kaushalesh Lal and Shampa Paul

What Drives China's Economy
Economic, Socio-Political, Historical and Cultural Factors
Qing-Ping Ma

Environmentally Sustainable Industrial Development in China
Yanqing Jiang and Xu Yuan

The China–US Trade War
Guoyong Liang and Haoyuan Ding

Analyzing Electoral Promises with Game Theory
Yasushi Asako

For more information about this series, please visit www.routledge.com/Routledge-Focus-on-Economics-and-Finance/book-series/RFEF

Analyzing Electoral Promises with Game Theory

Yasushi Asako

Routledge
Taylor & Francis Group

LONDON AND NEW YORK

First published 2021
by Routledge
2 Park Square, Milton Park, Abingdon, Oxon OX14 4RN

and by Routledge
52 Vanderbilt Avenue, New York, NY 10017

Routledge is an imprint of the Taylor & Francis Group, an informa business

British Library Cataloguing-in-Publication Data
A catalogue record for this book is available from the British Library

Library of Congress Cataloging-in-Publication Data
A catalog record has been requested for this book

ISBN: 978-0-367-44424-2 (hbk)
ISBN: 978-1-003-00963-4 (ebk)

Typeset in Times New Roman
by codeMantra

Contents

Contents

Figures

Table

Acknowledgements

This book is based on three of my papers: Asako (2015a), Asako (2015b), and Asako (2019). For these papers, I benefited from the comments of Swati Dhingra, Steven Durlauf, Amihai Glazer, Airo Hino, Yoichi Hizen, Hideki Konishi, Andrew Kydd, Ching-Yang Lin, Fumitoshi Moriya, John Morrow, Michael Munger, Yukihiro Nishimura, Tetsuro Okazaki, Daniel Quint, Ryoji Sawa, Ricardo Serrano-Padial, Orestis Troumpounis, Takashi Ui, Marek Weretka, and Mian Zhu. I am grateful for the financial support from the Nakajima Foundation and the JSPS KAKENHI (Grant Number 26780178, 17K13755, and 20K01734). I am also grateful to my family. My parents, Mitsunobu and Kiyoko, always supported me and encouraged me to achieve my dream. Without my wife Asako, I could not have gone through the journey of uncertainty to be a scholar. I also want to thank my sister, Naoko, and my family in Osaka: Fumiko, Akira, and Miyuki.

All the studies in this book were originally conducted when I was a Ph.D. student at the University of Wisconsin-Madison (UW-Madison), and some parts were originally included in my doctoral dissertation. The five years I spent at Madison were very exciting and valuable for me because I was fortunate enough to have met many people who supported me. If Bill Sandholm had not encouraged me to do the current study and supported me, I would never have been able to enjoy my life in UW-Madison. He trusted me from the very beginning, providing me with a high incentive to do my research. Scott Gehlbach always pointed me in the right direction. He was incredibly kind, and I will never forget that he cheered me up greatly when I was dejected. Marzena Rostek always motivated me; I found new directions from the discussions with her. I have been immensely fortunate to have such great advisors. Madison city itself is a great city that gave me many research ideas (even though there is nothing to do during winter). I dedicate this book to all in Madison and in the memory of Bill Sandholm.

1   Electoral promises in formal models

1.1 "Read my lips, no new taxes" and "end welfare as we know it"

Before an election, candidates announce platforms and the winner implements a policy after the election. Although politicians usually betray their platforms, such betrayal could prove costly. For example, in 1988 in the United States, George H. W. Bush promised, "read my lips, no new taxes." However, he increased taxes after becoming president. The media and voters noted this betrayal, and he lost the 1992 presidential election (Campbell, 2008, p. 104). On the contrary, in his 1992 campaign, Bill Clinton promised to "end welfare as we know it." In the 1994 midterm election, the Republican Party gained a majority of seats in the House of Representatives. Under pressure from Congress to keep his platform, he signed the welfare reform bill in 1996 (Weaver, 2000, Ch. 5).

These cases show two of the major characteristics of electoral promises. First, politicians decide policy on the basis of their platforms and the perceived cost of betrayal. If politicians betray their platforms, the people and media criticize them, they must address their electorate's complaints, their approval ratings may fall, and the possibility of them losing the next election might increase – as in the case of Bush.[1] Moreover, a stronger party or Congress may discipline such politicians, as in the case of Clinton.[2] On the basis of such costs of betrayal and the platform, the winner decides on the policy to be implemented after the election. The second characteristic is that politicians frequently prefer to use vague words and announce several policies in their electoral promises, a practice referred to as "political ambiguity." For instance, Clinton's promise above only stated his intention to reform the welfare system; he did not explicitly reveal the nature of the reform, which left the door open for further negotiation with the Republican Party.

However, most game-theoretical analyses of electoral competition have overlooked these two characteristics of electoral promises. Indeed, they avoid explicitly analyzing electoral platforms altogether. Formal models have supposed that candidates can choose only a single policy, meaning that political ambiguity never arises. Moreover, they fail to analyze platforms and policies separately and instead introduce two polar assumptions about platforms. On the one hand, models with *completely binding platforms* suppose that a politician cannot implement any policy other than the platform. That is, a politician never reneges on his/her promise and automatically implements the announced platform. The most famous example of such models is the electoral competition models in the Downsian tradition, that is, models based on those proposed by Downs (1957) and Wittman (1973). On the other hand, models with *non-binding platforms* suppose that a politician can implement any policy freely without any cost. In these models, voters do not believe electoral promises, so it is not meaningful to announce them. Therefore, a candidate's decisions on his/her platform are not explicitly analyzed. For example, this approach is taken in citizen-candidate models (Osborne and Slivinski, 1996; Besley and Coate, 1997) and retrospective voting models (Barro, 1973; Ferejohn, 1986; Besley, 2006). Neither model captures how, for example, Bush betrayed his platform and was then punished for doing so by the electorate or how Clinton kept his platform under pressure from Congress. To consider the effects of platforms on political competition, it is thus important to bridge these two settings; as Persson and Tabellini (2000) indicate, "(it) is thus somewhat schizophrenic to study either extreme: where platforms have no meaning or where they are all that matter. To bridge the two models is an important challenge" (p. 483).

The main purpose of this book is to provide a theoretical framework within which to analyze these two characteristics of campaign promises based on the electoral competition model in the Downsian tradition. Chapters 2 and 3 build a model with *partially binding platforms*, which supposes that although a candidate can choose any policy, there is a cost of betrayal. The policy to be implemented is affected by, but may be different from, the platform because this cost of betrayal increases with the degree of betrayal. As Figure 1.1 shows, models of completely binding platforms assume that candidates implement their platform, whereas models of non-binding platforms assume that candidates implement their own ideal policy. On the contrary, models of partially binding platforms suppose that candidates implement a policy situated between their platform and ideal policy. Chapter 4 discusses why candidates make vague promises in electoral competitions. Classically, the

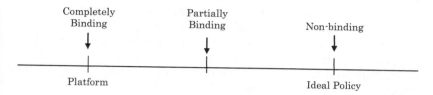

Figure 1.1 Completely, Non-, and Partially Binding Platforms.

In models of completely binding platforms, candidates implement their platform. In models of non-binding platforms, candidates implement their ideal policy. In models of partially binding platforms, candidates implement a policy between their platform and ideal policy.

convexity of a voter's preference is recognized as one possible reason driving political ambiguity. However, past studies have not shown it as an equilibrium. Based on the foregoing, I identify the conditions under which candidates choose an ambiguous platform in the equilibrium when voters have convex utilities. Chapters 2 to 4 are revised versions of Asako (2015a, 2015b, 2019), respectively. The proofs of all the propositions, lemmas, and corollaries are collected in the appendix.

1.2 Contributions and implications

The main contribution of this book is to show that the implications change considerably from past studies when electoral promises are explicitly analyzed in formal models. In particular, this book introduces two major characteristics of electoral promises, namely, partially binding platforms and political ambiguity, into the standard political competition model (Downs, 1957; Wittman, 1973). These new models can explain many aspects of real elections that cannot be predicted by previous frameworks. The findings confirm the importance of analyzing electoral promises explicitly using game theory.

The remainder of this chapter summarizes these new findings and explains how they differ from those of previous models.

1.2.1 *Two roles of electoral promises*

Because of the cost of betrayal, electoral promises serve as a commitment device *and* as a signal (see Chapters 2 and 3, respectively). First, because of the cost of betrayal, voters may believe that politicians will not betray their promise so severely to avoid paying this cost, and hence platforms can be considered as partial commitment devices to restrict a candidate's future policy choice. Second, because of the cost of betrayal, candidates do not have an incentive to choose a platform further away from their preferred policy. This is because the winner will betray the promise severely and pay large costs when his/her platform differs substantially from his/her ideal policy. Therefore, voters may be able to predict the position of the candidate's preferred policy through electoral promises. Thus, platforms can work as a signal about the candidate's policy preference. The following subsections summarize the main implications of each chapter.

1.2.2 Electoral promises as a commitment device

Chapter 2 mainly analyzes electoral promises as a commitment device by introducing the cost of betrayal into the Downsian electoral competition model with policy-motivated candidates. This model is generally known as Wittman's (1973) model. There is a one-dimensional policy space, and voters' preferred policies are distributed on this space. The preferred policies of 50% of voters are located to the left of the *median policy* and the remaining 50% of voters' preferred policies are located to the right. A voter who prefers the median policy is called the *median voter*.

One candidate prefers to implement a policy to the left of the median policy and the other candidate, to the right. Candidates announce their platforms before the election and the winner chooses a policy to be implemented thereafter. Politicians care about (1) the probability of winning, (2) the policy to be implemented, and (3) the cost of betrayal. I also analyze endogenous decisions to run on the basis of a simplified version of the citizen-candidate model (Osborne and Slivinski, 1996).

Partially binding platforms can explain the following observations from real elections, which cannot be explained by models of completely binding or non-binding platforms. Usually, in real elections, candidates have asymmetric characteristics (e.g., their policy preferences, the importance of policy, and costs of betrayal differ). Moreover, we frequently observe asymmetric outcomes (i.e., one candidate has a higher probability of winning than the other). Some candidates may avoid compromising on their principles to please voters and accept a lower probability of winning than their opponent, even though their probability of winning would be higher by compromising.

However, in existing frameworks, it is difficult to show asymmetric electoral outcomes. In models of completely binding platforms, both candidates propose the median voter's ideal policy regardless of their characteristics; hence, they have the same probability of winning (50%). In models of non-binding platforms, voters expect candidates' ideal policies to be implemented if they win. Then, only the candidate whose ideal policy is closer to the median policy can win, and no other characteristics affect the electoral outcome. On the contrary, in models of partially binding platforms, candidates with asymmetric characteristics can and will choose different platforms and policies to be implemented, since if their characteristics differ, one candidate may have a greater incentive to win – and would actually win – the election. This results in an asymmetric electoral outcome. Chapter 2 shows that an electoral outcome is asymmetric in the equilibrium when two candidates have different characteristics. For example, a more moderate candidate whose ideal policy is closer to the median policy wins against a more extreme candidate. Although this implication is the same as in models of non-binding platforms, this outcome is derived endogenously as opposed to exogenously as in those models. Similarly, if a candidate's cost of betrayal is higher than that of his/her opponent with the same degree of betrayal, the former candidate wins. If the cost is lower with the same degree of betrayal, a candidate will betray his/her platform more severely such that the realized cost of betrayal is higher; hence, this candidate has a lower incentive to win. In addition, a less policy-motivated candidate wins against a more policy-motivated candidate since the candidate with higher policy motivation will betray his/her platform more severely and pay his/her cost of betrayal.

In existing frameworks, it is also difficult to explain why a candidate runs even though he/she may lose in a two-candidate model. In models of completely binding platforms, both candidates have an equal probability of winning, and hence an explicit loser does not exist. In models of non-binding platforms, the winner will implement his/her ideal policy after the election, which means that the loser's decision to run does not affect the winner's policy. Thus, the loser has no reason to run. On the contrary, models of partially binding platforms show that even though a candidate is aware that he/she will lose, he/she may not deviate by withdrawing; therefore, he/she runs to induce his/her opponent to approach the median policy and thus the loser's ideal policy. For example, in the US 2010 primary elections of the Republican Party, one purpose of Tea Party-endorsed candidates was to induce Republican candidates or officeholders to be more conservative, which they succeeded in doing (Skocpol and Williamson, 2012). In obtaining this implication, the concept of partial bindingness is critical since the loser can change the winner's policy simply by entering the race.

1.2.3 Electoral promises as a signal

Chapter 3 analyzes the signaling role of electoral promises by introducing asymmetric information into the model presented in Chapter 2. That is, it assumes that a candidate's policy preferences are private information, meaning that voters are uncertain about them. A politician's preferences may change depending on the local conditions or important issues in an election. In particular, when candidates are not well known, it is difficult to know their preferences. Even in a political competition among parties, a party's preferences may change depending on the leader of the party. Voters are also uncertain of how power will be distributed between factions as a result of an intra-party negotiation after an election. If several parties form a coalition government, voters will be especially uncertain about the implemented policies, since these will depend on the negotiation among coalition members. For these reasons, it is reasonable to suppose that voters are uncertain about a candidate's policy preference.

The model in Chapter 3 simply supposes that each candidate is either moderate or extreme, with the moderate type's ideal policy closer to the median policy than that of the extreme type. A candidate knows his/her own type, whereas voters and the opponent do not. This extension provides a new implication to the model of electoral competition. Most studies of the two-candidate model of political competition show a politician's convergence toward the median policy. However, the patterns in real-world elections are frequently polarized. Indeed, in several countries, an extreme party has won an election by announcing a moderate platform, such as the Justice and Development Party of Turkey in 2007 (see Section 3.6 for more details).

One simple reason for this is polarized politicians. That is, politicians' policy preferences are extreme, which means these policies are located further from the median policy. Some candidates renege on their campaign promises to the electorate and implement an extreme policy after an election victory. However, most studies (including the results in Chapter 2) show that more moderate candidates beat more extreme candidates when they contest an election. Hence, winning politicians should not be so polarized. To understand polarization, it is thus important to answer the following questions. How do candidates with extreme policy preferences beat those whose policy preferences are less extreme? Why do extreme candidates have an incentive to win by compromising their preferences and approaching the median policy in an electoral campaign, but ultimately implementing a polarized, extreme policy after winning?

The striking result of the model presented in Chapter 3 is that an extreme candidate has a higher probability of winning than does a moderate candidate even though he/she will implement a more extreme policy. The important reason for the extreme candidate's higher probability of winning is that he/she has a stronger incentive to prevent his/her opponent from winning because his/her ideal policy is further from the opponent's policy than is that of a moderate candidate. My model describes this incentive for an extreme candidate by introducing two reasonable assumptions: partially binding platforms and uncertainty about a candidate's preference.

If voters believe in advance that a candidate is likely to be extreme, an extreme type has a higher probability of winning than a moderate type in a semi-separating equilibrium. In this equilibrium, an extreme type chooses a mixed strategy. With some probability, the extreme type announces the same platform as the moderate type, and with the remaining probability, approaches the median policy, thus revealing his/her type to voters. I call an extreme type a "pooling extreme type" when he/she imitates the moderate type, but a "separating extreme type" when he/she approaches the median policy.

A separating extreme type implements a more moderate policy than a pooling extreme type, but this is still a more extreme policy than that of a moderate type in the equilibrium, as Figure 1.2 shows. This is because an extreme type will betray his/her platform to a greater extent than a moderate type even though the platform of a separating extreme type is more moderate than that of a moderate type. While voters know a separating extreme type, they remain uncertain about the type of candidate who announces a moderate type's platform because this may be a moderate or a pooling extreme type candidate. Voters

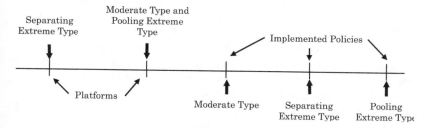

Figure 1.2 Platforms and Policies of Each Type.
This figure shows the platforms and policies chosen by each type (moderate type, separating extreme type, and pooling extreme type) of candidate whose ideal policy is to the right of the median policy.

around the median policy wish to avoid electing a pooling extreme type who will implement the most extreme policy. Thus, they forgo the chance to elect a moderate type who will implement the most moderate policy and choose a separating extreme type whose implemented policy lies between those of a moderate and a pooling extreme type. This is also why a separating extreme type can implement a more extreme policy than a moderate type but still defeat him/her. As a result, a separating extreme type has a higher probability of winning than a moderate type (and a pooling extreme type) in the equilibrium.

On the contrary, if voters believe that a candidate is likely to be a moderate type, a separating extreme type needs to rigorously approach the median policy to win. In this case, an extreme type simply imitates a moderate type with certainty (a pooling equilibrium). As a result, a moderate type can never have a higher probability of winning than an extreme type.

Combining the implications from Chapters 2 to 3, the following result is obtained. When voters know little about candidates and are uncertain about their policy preferences, an extreme candidate tends to win an election (Chapter 3). By contrast, when voters have sufficient information about candidates, a more moderate candidate tends to win (Chapter 2). Therefore, political polarization tends to occur when voters have insufficient information about candidates.

1.2.4 *Electoral promises with vague words*

Chapter 4 analyzes a possible cause of political ambiguity by allowing candidates to choose not only a single policy but also an ambiguous promise. A standard and classical interpretation of political ambiguity is a lottery (i.e., a probability distribution on single policies): candidates announce a lottery and voters choose the candidate who announces the better of the alternatives (Zeckhauser, 1969; Shepsle, 1972; Aragones and Postlewaite, 2002; Callander and Wilson, 2008).

One possible reason why candidates make such vague promises is because voters have convex utility functions. Zeckhauser (1969) was the first to interpret political ambiguity as a lottery, showing that the median policy can be defeated by a risky lottery when the voter's utility function is convex. Shepsle (1972) generalizes the findings of Zeckhauser (1969) and shows that a Condorcet winner does not exist when voters have convex utility functions. However, neither study establishes the existence of equilibria in which candidates announce ambiguous promises. Aragones and Postlewaite (2002) show political ambiguity as an equilibrium phenomenon using voters' convex utility functions. However, they assume that candidates need to provide a positive probability for their preferred policy. Thus, a campaign promise is always ambiguous when candidates commit to implementing a policy other than their own preferred policy. To the best of my knowledge, no existing studies show that a candidate chooses to make an ambiguous promise in the equilibrium because of the convex utility functions of voters, without any restriction on the candidate's choices.

Chapter 4 identifies the conditions under which candidates choose ambiguous promises in the equilibrium when voters have convex utility functions and candidates' choices are unrestricted. It extends the standard Downsian model with fully office-motivated candidates to allow a candidate to choose a lottery. Voters vote sincerely, and a candidate will implement a policy according to the probability distribution of the announced promise after he/she wins the election.

The findings are as follows. First, in a deterministic model without uncertainty, the unique Condorcet winner is the median policy when voters have concave or linear utility functions. However, no Condorcet winner exists when voters have convex utility functions. Therefore, two candidates choose the median policy in the equilibrium when voters have concave or linear utility functions, whereas no equilibrium exists in the case of convex utility functions. On the contrary, in a probabilistic voting model, where candidates are uncertain about voters' preferences, they choose ambiguous promises in the equilibrium when voters have convex utility functions and the distribution of voters' preferred policies is polarized. Therefore, for political ambiguity to be considered as an equilibrium phenomenon with convex utility functions, voters must be polarized and voting must be probabilistic.

Most prior studies assume that voters are risk-averse. However, there is no robust and clear evidence that voters have concave utility functions for all political issues. Osborne (1995) states, "I am uncomfortable with the implication of concavity that extremists are highly sensitive to differences between moderate candidates" (p. 275) and "it is not clear that evidence that people are risk-averse in economic decision-making has any relevance here" (p. 276). Furthermore, Kamada and Kojima (2014) state that "(e)conomic policy is arguably a concave issue, given the evidence that individuals are risk-averse in financial decisions. By contrast, voters may have convex utility functions on moral or religious issues" (p. 204). They claim that an ambiguous promise tends to be used for non-economic issues, which may be a convex issue.

For example, suppose a voter who is conservative and is against introducing regulation on gun ownership. Thus, this voter's preferred policy is no regulation on gun ownership and his/her ideal policy can be located at the right end when the policy space represents the degree of regulation on gun ownership, as shown in Figure 1.3. When this voter has concave preferences as in Figure 1.3(a), it means that his/her utility decreases little when a weak regulation is introduced, whereas he/she cares a lot about a small difference in strict regulations since his/her utility changes greatly with a marginal difference. On the contrary, when this voter has convex preferences as in Figure 1.3(b), a small change in strict regulations changes his/her utility little since it is already very low. Rather, the utility of this voter decreases considerably by introducing any type of regulation – even a weak one. To represent the conservative voter's preference, the latter should thus be more appropriate than the former.

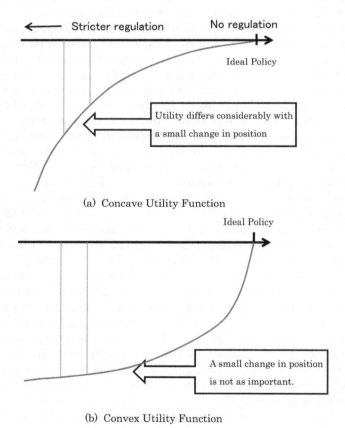

(a) Concave Utility Function

(b) Convex Utility Function

Figure 1.3 Concave and Convex Preferences.

Shepsle (1972) states the following:

> In the 1968 presidential campaign, both Nixon's "I have a plan" statements on the Vietnam issue and Humphrey's "law and order with justice" slogan on "the social issue" suggest that equivocal pronouncements during the course of campaign are a common and recurrent theme in American electoral politics.
>
> (p. 555)

These are examples of ambiguity regarding non-economic issues, and public opinion on the Vietnam war was almost equally divided and polarized between pro-escalation and anti-escalation (Verba et al., 1967). Therefore, the model in Chapter 4 provides one possible explanation for why Nixon chose an ambiguous promise.

1.2.5 Summary

Table 1.1 highlights the contributions and implications of each chapter. This table shows the new settings and novel implications that extend the findings of past studies. To clarify the novelty of my implications, the next section summarizes previous studies and explains how they differ from my model.

Table 1.1 New Settings and Implications of Each Chapter

	New Setting(s)	*New Implication(s)*
Chapter 2	• Partially binding platform	• Asymmetric electoral outcome • The reason why a loser runs an election
Chapter 3	• Partially binding platform • Uncertainty about a candidate's policy preference	• Political polarization (the reason why an extreme candidate wins an election)
Chapter 4	• Allowing candidates to choose a lottery as a promise • Uncertainty about voters' policy preference	• Political ambiguity with a convex policy preference

1.3 Related studies

1.3.1 Electoral promises as a commitment device

Few previous studies consider platforms as a partial commitment device by introducing the cost of betrayal as in Chapter 2.
Austen-Smith and Banks (1989) consider a two-period game based
on a retrospective voting model in which the probability of winning
in the next election decreases if office-motivated candidates betray
the platform. Grossman and Helpman (2005, 2008) develop a legislative model in which office-motivated parties announce platforms
before an election and the victorious legislators who are policy-
motivated decide policy. If legislators betray the party platform, the
party punishes them. Hummel (2010) supposes that there exist costs
when policy announcements are different between primary and general elections. On the contrary, my model is based on the prospective
voting and two-candidate competition models without primaries
and assumes that policy-motivated candidates decide on both a platform and a policy.

Note that Austen-Smith and Banks (1989) consider only a decrease
in the probability of winning as the cost of betrayal, while Grossman
and Helpman (2005, 2008) consider only a party's discipline as the cost
of betrayal. However, the cost of betrayal in this book also includes
many types of costs such as a decrease in approval ratings and a negotiation cost with Congress; therefore, I include these in the current
term as the cost of betrayal.

1.3.2 Electoral promises as a signal

Banks (1990) and Callander and Wilkie (2007) consider campaign
platforms as signals about the candidate's policy preference by introducing the cost of betrayal as in Chapter 3. However, there are two important differences between my models and theirs. First, in their work,
candidates automatically implement their own ideal policies after
winning an election. However, if there is a cost of betrayal, a rational
candidate would wish to adjust the policy to be implemented to reduce
that cost after the election. Second, Banks (1990) and Callander and
Wilkie (2007) consider that candidates care about policy only when
they win – their utility is set to zero when they lose regardless of the
policy implemented by their opponent. However, policy-motivated
candidates should care about policy when they lose.

I relax these assumptions and make more reasonable ones by examining rational choices about a policy to be implemented and candidates who care about policy regardless of the election results. These two differences are critical to obtaining my results. For example, Chapter 2 shows that some candidates have an incentive to run even though they know *with certainty* that they will lose an election because it induces the opponent to approach the median policy, and thus the loser's ideal policy. It is impossible to obtain this implication by assuming the above two assumptions. First, if candidates implement their own ideal policies automatically, it is impossible to induce the opponent to approach the median policy. Thus, for the loser, there is no way to change the opponent's policy. Second, if a candidate does not care about policy when he/she loses, the loser does not have an incentive to change the opponent's policy because his/her utility is zero in all cases. Thus, these two differences provide a way and an incentive to induce the opponent to approach the median policy.

Banks (1990) and Callander and Wilkie (2007) show that a moderate type may defeat an extreme type when asymmetric information about each candidate's ideal policies exists. On the contrary, the results in Chapter 3 contradict those in the aforementioned studies. That is, an extreme type has a higher probability of winning than a moderate type. It is also impossible to gain this opposite result by assuming the above two assumptions. First, if candidates implement their own ideal policies automatically, an extreme candidate will lose against a moderate type when he/she reveals his/her type to voters. Thus, a separating extreme type cannot obtain a higher probability of winning in a semi-separating equilibrium. Second, if a candidate does not care about policy when he/she loses, an extreme type will not have as strong an incentive to prevent his/her opponent from winning. Therefore, under the assumptions of Banks (1990) and Callander and Wilkie (2007), an extreme type does not have either a method or an incentive to win against a moderate type.

In Huang (2010), candidates strategically choose both a platform and an implemented policy but do not care about policy when they lose. Huang (2010) also supposes sufficiently large benefits from holding office and shows that candidates cluster around or at the median policy. On the contrary, my models in Chapters 2 and 3 suppose that candidates care about policy even after losing and does not consider the benefits from holding office as large. Other research also considers that an electoral promise can work as a signal of the functioning of the economy (Schulz, 1996), the candidate's degree of honesty (Kartik and McAfee, 2007), and political motivation (Callander, 2008).

1.3.3 Endogenous candidates

Osborne and Slivinski (1996) explain why some candidates run for election even though they might lose. They assume non-binding platforms; hence, a candidate cannot induce his/her opponent to compromise more because the policy to be implemented is given as an ideal policy. Thus, the reason a candidate runs to lose is different in their study and my model. According to Osborne and Slivinski (1996), the loser runs to change the identity of the winner (i.e., decrease an undesirable candidate's probability of winning). On the contrary, in the model presented in Chapter 2, the loser runs to induce the winner to approach the loser's ideal policy. Moreover, in Osborne and Slivinski (1996), a candidate who runs and is certain to lose never appears in a two-candidate competition. Wada (1996, Ch. 2) also shows that some candidates may run even though they will lose using a different model framework. Ishihara's (2020) result using a repeated two-candidate competition model is similar to that of my model, which analyzes a one-shot game.

1.3.4 Valence

The models in Chapters 2 and 3 also relate to models of "valence" since they show asymmetric electoral outcomes in the equilibrium. Several studies such as Ansolabehere and Snyder (2000), Aragones and Palfrey (2002), Groseclose (2001), Kartik and McAfee (2007), and Callander (2008) consider the effects of a candidate's character or personality, as indicated by Stokes (1963) as valence, and show an asymmetric probability of winning in a political competition. These studies assume that there are advantaged and disadvantaged candidates, and an advantage is given exogenously as valence. Voters care not only about policy but also about valence, suggesting that an advantaged candidate with good valence has a higher probability of winning an election. Therefore, an extreme candidate may win against a moderate candidate if he/she has good valence, as shown by Kartik and McAfee (2007) and Callander (2008).[3] On the contrary, I derive an asymmetric probability of winning endogenously in Chapter 2 and show that an extreme candidate wins against a moderate candidate without such exogenous valence in Chapter 3.

1.3.5 Political ambiguity

1.3.5.1 Causes of political ambiguity

Prior studies use formal models to indicate various mechanisms that generate political ambiguity. There are two main types of models:

voter-centered and candidate-centered models. Voter-centered models suppose that voters prefer a higher degree of ambiguity and that candidates choose an ambiguous promise to win an election. This category includes models in which voters have convex utility functions. Callander and Wilson (2008) indicate that voters may develop a taste for ambiguity using the notion of context-dependent voting, as introduced by Callander and Wilson (2006).[4] Kartik et al. (2017) suppose that voters are uncertain about their own ideal policy and that only politicians receive policy-relevant information after an election. In this case, voters prefer a politician who announces a vague promise when he/she shares voters' policy preferences, because they prefer to allow the politician the discretion to adopt policies. Moreover, if voters are boundedly rational, candidates who make ambiguous promises may win the election. For example, voters believe incorrectly that their favored candidate's position is closer to their preferred policy than it actually is (Jensen, 2009), voters are not expected utility maximizers and have Knightian uncertainty (Berliant and Konishi, 2005), or voters take at face value the campaign promise announced by candidates without being able to decipher their strategy (Demange and Van der Straeten, 2017; Szembrot, 2017).

On the contrary, candidate-centered models suppose that voters dislike ambiguity, but that candidates prefer ambiguity. First, candidates may prefer political ambiguity because of its direct (non-electoral) benefits. For example, they may not know which policy is most expedient (Aragones and Neeman, 2000) and may want the flexibility to implement their own preferred policy (Alesina and Cukierman, 1990). Furthermore, a party may be able to recruit a greater number of elites by allowing for ideological diversity (Jensen and Lee, 2017). Second, candidates may be able to obtain indirect (electoral) benefits from political ambiguity even though voters prefer a less ambiguous policy. For example, when candidates are uncertain about the position of the median policy, they may prefer to maintain ambiguity (Glazer, 1990). This is especially true in primary elections because candidates have less information about voters' preferences (Meirowitz, 2005). Moreover, if campaign platforms are decided sequentially, the follower, who makes policy decisions later than his/her opponent, has a significant advantage when a Condorcet winner does not exist. As a result, candidates prefer to retain political ambiguity (Kamada and Sugaya, 2020).

1.3.5.2 Definitions of political ambiguity

This book interprets political ambiguity as a lottery that includes several policies; however, it has alternative definitions. Some studies interpret a set of policies as an ambiguous policy and do not consider a candidate's decision-making on a probability distribution in this set (Glazer, 1990; Aragones and Neeman, 2000; Jensen, 2009; Kartik et al., 2017; Kamada and Sugaya, 2020). Meirowitz (2005), Berliant and Konishi (2005), and Szembrot (2017) suppose that ambiguity exists when candidates do not announce anything in their campaigns.[5] Most of these studies suppose that candidates do not have the discretion to decide the probability distribution on policies, which is given exogenously in the model. That is, candidates cannot change the degree of ambiguity completely. On the contrary, studies that suppose political ambiguity as a lottery assume that candidates can choose any probability distribution freely. The reality should be between these two definitions; candidates have some (but not perfect) discretion to decide a probability distribution. However, I suppose political ambiguity as a lottery in Chapter 4 to investigate the strategic choices of candidates on the degree of ambiguity.

There are two justifications to suppose that candidates announce a specific probability distribution on policies in their campaign. First, using words, candidates may induce voters to have specific expectations. For example, in Japan, Prime Minister Shinzo Abe made an ambiguous announcement stating that he will increase the consumption tax rate in 2019 if a severe financial crisis (or a similar event) does not occur. Thus, voters may think that although the probability of such an event is high since financial crises do not occur frequently, it is not 100% certain. (If COVID-19 was spreading in early 2019, it might not be increased.) Second, candidates may induce voters to have specific expectations by allocating weights to each policy (Page, 1976).[6] Voters believe that a policy with a higher weight is more likely to be implemented. These interpretations suppose a probability distribution of an ambiguous promise as the beliefs held by voters. In other words, a candidate can use weights or words to induce voters to form specific beliefs.

1.4 Future work

The Downsian electoral competition model is applied to many other topics such as special interest politics, media, authoritarian politics, public policy, and macroeconomics. As this book shows, explicit analyses of electoral platforms induce many predictions that differ from those of previous studies. Therefore, applications of the models presented in this book should be an interesting subject for future research to provide new implications. Moreover, not only the Downsian model but also many other formal models such as the political agency model can be used to analyze electoral platforms.

The other possible area of future research is to endogenize the cost of betrayal. In this book, the cost of betrayal simply depends on the degree of betrayal; however, it may also be decided endogenously. For example, one type of cost of betrayal is a decrease in the probability of winning the next election. To analyze such reputational costs, a dynamic model comprising two or more periods should be analyzed. Moreover, depending on the economic and social situation, the cost of betrayal changes before and after an election. For example, even if a politician promises "no new taxes," voters may allow politicians to betray their platforms by changing taxes after a natural disaster occurs.

In Chapter 3, the model of partially binding platforms with asymmetric information assumes that candidates are symmetric and that only two types of candidates exist. However, in reality, these characteristics may differ, as shown by the model in Chapter 2. The model in Chapter 3 should also be extended to analyze asymmetric candidates, which may lead to different implications from those in Chapter 2.

This book analyzes the case with two candidates and a one-dimensional policy space. It may be able to be extended to analyze the case with multiple candidates and/or a multidimensional policy space. Moreover, the generalization to a continuous policy space in Chapter 4 would be another important extension. If a continuous policy space is supposed, it would also be possible to analyze a more realistic case in which utilities are neither always convex nor always concave.

In the past, formal models of political competition did not analyze electoral promises explicitly even though they are crucial aspects of real elections. This book shows that it is possible to obtain more important implications by analyzing them explicitly. Promising future work in formal political theory should include analyzing electoral platforms explicitly even though this "promise" is non-binding.

Notes

1 Some studies show the relationship between the media and the credible commitment of politicians. For example, Reinikka and Svensson (2005) study a newspaper campaign in Uganda aimed at reducing the capture of public funds by providing schools (parents) with information to monitor local officials' handling of a large education grant program, and they show that it actually reduced corruption. Djakov et al. (2003) empirically show that policymaking is distorted if the media is owned by the government.

2 Cox and McCubbins (1994) and Aldrich (1995) emphasize this point from the historical aspects of American parties. Snyder and Groseclose (2000)

and McCarty et al. (2001) empirically show that there are various party discipline in the US Congress. McGillivray (1997) compares high with low discipline in trade policies.

3 Kartik and McAfee (2007) provide good "character" exogenously and interpret it as integrity. Callander (2008) provides this valence endogenously using policy motivation, whereas a candidate's policy motivation is given exogenously. In both studies, more extreme policy can be a signal of good valence.

4 Context-dependent voting means that voters are interested not only in the policies of the party in question, but also in the relative attractiveness of the opposition's policies.

5 On the contrary, Alesina and Cukierman (1990), Jensen and Lee (2017), and Demange and Van der Straeten (2017) define the level of ambiguity as the variance in the noise of the policy outcomes observed by voters.

6 Page (1976) considers that political ambiguity arises when candidates allocate their limited resources (emphasis) among several policies. If candidates do not allocate sufficient resources to a policy, its promise to voters becomes vague. Chappell (1994) and Dellas and Koubi (1994) follow a similar interpretation.

Bibliography

Aldrich, J., 1995, *Why Parties? The Origin and Transformation of Political Parties in America*, Chicago: University of Chicago Press.

Alesina, A., and A. Cukierman, 1990, "The Politics of Ambiguity," *Quarterly Journal of Economics* 4, pp. 829–850. (https://doi.org/10.2307/2937875)

Ansolabehere, S., and J. Snyder Jr., 2000, "Valence Politics and Equilibrium in Spatial Election Models," *Public Choice* 103, pp. 327–336. (https://doi.org/10.1023/A:1005020606153)

Aragones, E., and Z. Neeman, 2000, "Strategic Ambiguity in Electoral Competition," *Journal of Theoretical Politics* 12(2), pp. 183–204. (https://doi.org/10.1177/0951692800012002003)

Aragones, E., and T. Palfrey, 2002, "Mixed Equilibrium in a Downsian Model with a Favored Candidate," *Journal of Economic Theory* 103(1), pp. 131–161. (https://doi.org/10.1006/jeth.2001.2821)

Aragones, E., and A. Postlewaite, 2002, "Ambiguity in Election Games," *Review of Economic Design* 7, pp. 233–255. (https://doi.org/10.1007/s100580200081)

Asako, Y., 2015a, "Partially Binding Platforms: Campaign Promises vis-a-vis Cost of Betrayal," *Japanese Economic Review* 66(3), pp. 322–353. (https://doi.org/10.1111/jere.12053)

Asako, Y., 2015b, "Campaign Promises as an Imperfect Signal: How Does an Extreme Candidate Win against a Moderate One?," *Journal of Theoretical Politics* 27(4), pp. 613–649. (https://doi.org/10.1177/0951629814559724)

Asako, Y., 2019, "Strategic Ambiguity with Probabilistic Voting," *Journal of Theoretical Politics* 31(4), pp. 626–641. (https://doi.org/10.1177/0951629819875516)

Austen-Smith, D., and J. Banks, 1989, "Electoral Accountability and Incumbency," in P. Ordeshook, ed., *Models of Strategic Choice in Politics*, Ann Arbor: University of Michigan Press, pp. 121–150.

Banks, J., 1990, "A Model of Electoral Competition with Incomplete Information," *Journal of Economic Theory* 50(2), pp. 309–325. (https://doi.org/10.1016/0022-0531(90)90005-5)

Barro, R., 1973, "The Control of Politicians: An Economic Model," *Public Choice* 14(1), pp. 19–42. (https://doi.org/10.1007/BF01718440)

Berliant, M., and H. Konishi, 2005, "Salience: Agenda Choices by Competing Candidates," *Public Choice* 125, pp. 129–149. (https://doi.org/10.1007/s11127-005-3412-9)

Besley, T., 2006, *Principled Agents?: The Political Economy of Good Government*, Oxford: Oxford University Press.

Besley, T., and S. Coate, 1997, "An Economic Model of Representative Democracy," *Quarterly Journal of Economics* 112(1), pp. 85–114. (https://doi.org/10.1162/003355397555136)

Callander, S., 2008, "Political Motivations," *Review of Economic Studies* 75(3), pp. 671–697. (https://doi.org/10.1111/j.1467-937X.2008.00488.x)

Callander, S., and S. Wilkie, 2007, "Lies, Damned Lies and Political Campaigns," *Games and Economic Behavior* 60(2), pp. 262–286. (https://doi.org/10.1016/j.geb.2006.12.003)

Callander, S., and C. Wilson, 2006, "Context-Dependent Voting," *Quarterly Journal of Political Science* 1(3), pp. 227–254. (http://dx.doi.org/10.1561/100.00000007)

Callander, S., and C. Wilson, 2008, "Context-Dependent Voting and Political Ambiguity," *Journal of Public Economics* 92(3–4), pp. 565–581. (https://doi.org/10.1016/j.jpubeco.2007.09.002)

Campbell, J., 2008, *The American Campaign: U.S. Presidential Campaign and the National Voter*, 2nd Edition, College Station: Texas A&M University Press.

Chappell. H. W. Jr., 1994, "Campaign Advertising and Political Ambiguity," *Public Choice* 79, pp. 281–303. (https://doi.org/10.1007/BF01047774)

Cox, G., and M. McCubbins, 1994, *Legislative Leviathan: Party Government in the House*, Los Angeles: University of California Press.

Dellas, H., and V. Koubi, 1994, "Smoke Screen: A Theoretical Framework," *Public Choice* 78, pp. 351–358. (https://doi.org/10.1007/BF01047763)

Demange, G., and K. Van der Straeten, 2017, "Communicating on Electoral Platforms," *Journal of Economic Behavior & Organization*, forthcoming. (https://doi.org/10.1016/j.jebo.2017.03.006)

Djankov, S., C. McLiesh, T. Nenova, and A. Shleifer, 2003, "Who Owns the Media?," *Journal of Law and Economics* 46(2), pp. 341–382. (https://doi.org/10.1086/377116)

Downs, A, 1957, *An Economic Theory of Democracy*, New York: Harper and Row.

Ferejohn, J., 1986, "Incumbent Performance and Electoral Control," *Public Choice* 50, pp. 5–26. (https://doi.org/10.1007/BF00124924)

Glazer, A., 1990, "The Strategy of Candidate Ambiguity," *American Political Science Review* 84(1), pp. 237–241. (https://doi.org/10.2307/1963640)

Groceclose, T., 2001, "A Model of Candidate Location When One Candidate Has a Valence Advantage," *American Journal of Political Science* 45(4), pp. 862–886. (https://doi.org/10.2307/2669329)

Grossman, G., and E. Helpman, 2005, "A Protectionist Bias in Majoritarian Politics," *The Quarterly Journal of Economics* 120(4), pp. 1239–1282. (https://doi.org/10.1162/003355305775097498)

Grossman, G., and E. Helpman, 2008, "Party Discipline and Pork-Barrel Politics," in E. Helpman, ed., *Institutions and Economic Performance*, Cambridge: Harvard University Press, pp. 329–360.

Huang, H., 2010, "Electoral Competition when Some Candidates Lie and Others Pander," *Journal of Theoretical Politics* 22(3), pp. 333–358. (https://doi.org/10.1177/0951629810365151)

Hummel, P., 2010, "Flip-Flopping from Primaries to General Elections," *Journal of Public Economics* 94, pp. 1020–1027. (https://doi.org/10.1016/j.jpubeco.2010.08.006)

Ishihara, A., 2020, "Strategic Candidacy for Political Compromise in Party Politics," *Journal of Theoretical Politics*, forthcoming.

Jensen, C., and D. Lee, 2017, "Predicting Ambiguity: Costs, Benefits, and Party Competition," *Political Research Quarterly* 70(2), pp. 301–313. (https://doi.org/10.1177/1065912917691139)

Jensen, T., 2009, "Projection Effects and Strategic Ambiguity in Electoral Competition," *Public Choice* 141, pp. 213–232. (https://doi.org/10.1007/s11127-009-9449-4)

Kamada, Y., and F. Kojima, 2014, "Voter Preferences, Polarization, and Electoral Policies," *American Economic Journal: Microeconomics* 6(4), pp. 203–236. (https://doi.org/10.1257/mic.6.4.203)

Kamada, Y., and T. Sugaya, 2020, "Optimal Timing of Policy Announcements in Dynamic Election Campaigns," *Quarterly Journal of Economics*, forthcoming.

Kartik, N., and P. McAfee, 2007, "Signaling Character in Electoral Competition," *The American Economic Review* 97(3), pp. 852–870. (https://doi.org/10.1257/aer.97.3.852)

Kartik, N., R. Van Weelden, and S. Wolton, 2017, "Electoral Ambiguity and Political Representation," *American Journal of Political Science* 61(4), pp. 958–970. (https://doi.org/10.1111/ajps.12310)

McCarty N., K. Poole, and H. Rosenthal, 2001, "The Hunt for Party Discipline in Congress," *American Political Science Review* 95(3), pp. 673–687. (https://doi.org/10.1017/S0003055401003069)

McGillivray, F., 1997, "Party Discipline as a Determinant of the Endogenous Formation of Tariffs," *American Journal of Political Science* 41(2), pp. 584–607. (https://doi.org/10.2307/2111778)

Meirowitz, A., 2005, "Informational Party Primaries and Strategic Ambiguity," *Journal of Theoretical Politics* 17(1), pp. 107–136. (https://doi.org/10.1177/0951629805047800)

Osborne, M. J., 1995, "Spatial Models of Political Competition under Plurality Rule: A Survey of Some Explanations of the Number of Candidates and the Positions They Take," *Canadian Journal of Economics* 28(2), pp. 261–301. (https://doi.org/10.2307/136033)

Osborne, M., and A. Slivinski, 1996, "A Model of Political Competition with Citizen-Candidates," *The Quarterly Journal of Economics* 111(1), pp. 65–96. (https://doi.org/10.2307/2946658)

Page, B. I., 1976, "The Theory of Political Ambiguity," *The American Political Science Review* 70, pp. 742–752. (https://doi.org/10.2307/1959865)

Persson, T., and G. Tabellini, 2000, *Political Economics: Explaining Economic Policy*, Cambridge: The MIT Press.

Reinikka, R., and J. Svensson, 2005, "Fighting Corruption to Improve Schooling: Evidence from a Newspaper Campaign in Uganda," *Journal of the European Economic Association* 3(2–3), pp. 259–267. (https://doi.org/10.1162/jeea.2005.3.2-3.259)

Schultz, C., 1996, "Polarization and Inefficient Policies," *Review of Economic Studies* 63(2), pp. 331–344. (https://doi.org/10.2307/2297855)

Shepsle, K., 1972, "The Strategy of Ambiguity: Uncertainty and Electoral Competition," *American Political Science Review* 66, pp. 555–568. (https://doi.org/10.2307/1957799)

Skocpol, T., and V. Williamson, 2012, *The Tea Party and the Remaking Republican Conservatism*, New York: Oxford University Press.

Snyder, J., and T. Groseclose, 2000, "Estimating Party Influence in Congressional Role-Call Voting," *American Journal of Political Science* 44(2), pp. 193–211. (https://doi.org/10.2307/2669305)

Stokes, D., 1963, "Spatial Models of Party Competition," *American Political Science Review* 57(2), pp. 368–377. (https://doi.org/10.2307/1952828)

Szembrot, N., 2017, "Are Voters Cursed When Politicians Conceal Policy Preference?," *Public Choice* 173, pp. 25–41. (https://doi.org/10.1007/s11127-017-0461-9)

Verba, S., R. A. Brody, E. B. Parker, N. H. Nie, N. W. Polsby, P. Ekman, and G. S. Black, 1967, "Public Opinion and the War in Vietnam," *American Political Science Review* 61, pp. 317–333. (https://doi.org/10.2307/1953248)

Wada, J., 1996, *The Japanese Election System: Three Analytical Perspectives*, London: Routledge.

Weaver, K., 2000, *Ending Welfare as We Know It*, Brookings: Brookings Institution Press.

Wittman, D., 1973, "Parties as Utility Maximizers," *American Political Science Review* 67(2), pp. 490–498. (https://doi.org/10.2307/1958779)

Zeckhauser, R., 1969, "Majority Rule with Lotteries on Alternatives," *The Quarterly Journal of Economics* 83(4), pp. 696–703. (https://doi.org/10.2307/1885458)

2 Electoral promises as a commitment device

2.1 Introduction

This chapter extends the basic political-competition model in the Downsian tradition (Downs, 1957) by considering partially binding platforms, which suppose that although a candidate can choose any policy, there is a cost for betrayal.[1] The policy to be implemented is affected by, but may be different from, the platform because of this cost, which increases with the degree of betrayal. In the model specified in this chapter, it is assumed all players have complete information; therefore, there no uncertainty.

The model of partially binding platforms presents the following two implications. First, although it has been difficult to show asymmetric electoral outcomes in previous frameworks, the model of partially binding platforms can show that candidates with asymmetric characteristics can and will choose different platforms and policies to be implemented. This is because, if their characteristics differ, one candidate may have a greater incentive to win – and would actually win – the election. As a result, an electoral outcome is asymmetric in equilibrium when two candidates have different characteristics.

Second, in existing frameworks, it has been difficult to explain why a candidate runs for an election even though he/she may lose in a two-candidate model. By contrast, the model of partially binding platforms shows that even though a candidate is aware that he/she will lose, he/she may not deviate by withdrawing and runs in order to induce the opponent to approach the median policy and thus, the loser's ideal policy.

2.2 The model

2.2.1 Setting

The policy space is $\mathbb{R}$. There is a continuum of voters, and their ideal policies are distributed on some interval of $\mathbb{R}$. This distribution function is continuous and strictly increasing, which means there exists a unique median voter's ideal policy (the median policy), x_m. Assume that this distribution is symmetric and single-peaked about x_m.

Suppose there are two potential candidates, and each decides whether to run for office.[2] Denote x_i as the ideal policy of potential candidate (or voter) i. If a candidate wins, he/she will obtain a benefit from holding office, $b > 0$, which is not related to the ideal policy. However, candidates do have to pay a cost for running, $k > 0$.

In the second period, each candidate announces a platform, denoted by $z_i \in \mathbb{R}$. On observing the available platforms, voters can ascertain correctly the policy to be implemented by each candidate as they have complete information. Based on the (expected) policy to be implemented, all voters cast their votes according to the plurality rule; that is, the candidate with the most votes wins. Note that voting is sincere, and I rule out weakly dominated voting strategies. In the last period, the winning candidate, i, decides on the actual policy to be implemented, denoted by χ_i.

The voter and candidate experience a disutility if the implemented policy differs from their ideal policy. In line with Calvert's (1985) study, this disutility is represented by $-\beta u(|\chi - x_i|)$, where χ represents the policy implemented by the winner. Assume that $u(.)$ satisfies $u(0) = 0$, $u'(d) > 0$, and $u''(d) \geq 0$ when $d > 0$. The level of political motivation is $\beta \in (0, \infty)$, in which a higher or lower β means a candidate is more policy motivated or more office motivated, respectively. Without loss of generality, for now, I assume $\beta = 1$ for both candidates. I discuss the case in which candidates have different β values later.

If the implemented policy is not the same as that of the platform, the winning candidate incurs a cost of betrayal. The function describing this cost is $\lambda c(|z_i - \chi_i|)$. Assume that $c(.)$ satisfies $c(0) = 0$, $c'(0) = 0$, $c'(d) > 0$, and $c''(d) > 0$ when $d > 0$. Here, $\lambda > 0$ represents the relative importance of betrayal. In the last period, the winning candidate chooses a policy that maximizes $-u(|\chi - x_i|) - \lambda c(|z_i - \chi|)$. Denote $\chi_i(z_i) = \operatorname{argmax}_\chi \left[-u(|\chi - x_i|) - \lambda c(|z_i - \chi|) \right]$. Therefore, if the candidate runs and wins, the utility is $-u\big(\|\chi_i(z_i)-x_i\|\big)-\lambda c\big(\|z_i-\chi_i(z_i)\|\big)+b-k$. If the candidate runs but loses, the utility is $-u\big(\|\chi-x_i\|\big)-k$. I assume $b>k$. In other words, potential candidates have an incentive to run if they will definitely win by announcing their ideal policy as their platform (i.e., $z_i=x_i=\chi_i(z_i)$ and $-u\big(\|x_i-x_i\|\big)-\lambda c\big(\|x_i-x_i\|\big)=0$). Furthermore, assume that if no candidate enters the election, all obtain a payoff of $-\infty$, as in Osborne and Slivinski's (1996) study. As I assume $b>k$, even if a status-quo policy is introduced, at least one candidate will enter the race. Hence, the position of a status-quo policy does not matter.

The equilibrium concept is a subgame perfect equilibrium. I restrict the analysis to a pure strategy equilibrium. I also concentrate on the typical case in which one candidate's ideal policy is to the left of the median policy, x_m, while that of the other candidate is to the right. Here, the candidate whose ideal policy is to the left of the median policy is denoted as candidate L, and the other is candidate R (i.e., $x_L<x_m<x_R$). In summary, the timing of events is as follows.

1   Two potential candidates decide whether to run. If no candidate enters the election, all voters and potential candidates obtain a payoff of $-\infty$.
2   The candidates who decide to run announce their platforms.
3   Voters vote. The candidate with the most votes wins. If only one candidate runs, this candidate wins with a probability of 1.
4   The winning candidate chooses the policy to be implemented.

2.2.2 Policy implemented by the winner

First, the scenario after period 2, that is, the no-entry model in which the two potential candidates have already decided to run, is analyzed. I ignore the cost of running because it is a sunk cost at this stage.

In the last period, the winning candidate implements the policy that maximizes the utility after a win, $-u\big(\|\chi_i(z_i)-x_i\|\big)-\lambda c\big(\|z_i-\chi_i(z_i)\|\big)$, given z_i.

Lemma 2.1

Consider that $u''(d)>0$, for any $d>0$. In equilibrium, $\chi_i(z_i)$ satisfies

$$\lambda=\frac{u'\big(\|\chi_i(z_i)-x_i\|\big)}{c'\big(\|z_i-\chi_i(z_i)\|\big)} \tag{2.1}$$

given $z_i \neq x_i$. If λ goes to infinity, $\chi_i(z_i)$ converges to z_i. If λ goes to zero, $\chi_i(z_i)$ converges to x_i.

The policy to be implemented will lie somewhere between the platform policy and the ideal policy, as shown in Figure 1.1 in Chapter 1. When λ increases, the policy that the winning candidate chooses to implement approaches the platform policy. Similarly, when λ decreases, the implemented policy approaches the ideal policy. If the policy a candidate chooses to implement lies closer to the median policy than that of the opponent, this candidate is certain to win.

There are three additional implications. First, if the disutility function is linear, given platform z_i, the winner may prefer to implement x_i rather than $\chi_i(z_i)$, which satisfies (2.1). Here, I denote $u'(d) = \bar{u} > 0$, which is constant for all $d \geq 0$ because $u(.)$ is a linear function.

Corollary 2.2

Consider that $u''(d) = 0$, for all $d \geq 0$. Then, given $z_i \neq x_i$, if λ is sufficiently low such that $\lambda < \bar{u} / c'(|z_i - x_i|)$, the winner implements x_i. Otherwise, the winner implements $\chi_i(z_i)$, which satisfies (2.1).

Proof: See Appendix 2.A.1.

However, a candidate never chooses $z_i \neq x_i$ and $\chi_i(z_i) = x_i$ in equilibrium. This is because, in committing to x_i, it is better to choose $z_i = \chi_i(z_i) = x_i$, as there is then no need to pay the cost of betrayal. Thus, if a decision on z_i is included in the analyses, then in equilibrium, a candidate will either choose $\chi_i(z_i)$, which satisfies (2.1), or $z_i = \chi_i(z_i) = x_i$, as Corollary 2.7 will show. Thus, this boundary case is trivial.

Second, if a candidate's platform approaches his/her own ideal policy, the cost of betrayal and the disutility from winning decreases. In other words, if a candidate compromises more toward the median voter, his/her expected utility from winning decreases.

Corollary 2.3

As z_i approaches x_i, $u(|\chi_i(z_i) - x_i|)$ and $c(|z_i - \chi_i(z_i)|)$ decrease.

Proof: See Appendix 2.A.2.

Third, if the benefit from holding office b is very large, candidates are less concerned about the cost of betrayal, and hence, the policies they choose to implement converge to the median policy. That is, both candidates will implement the median policy, as in the basic Downsian model. Denote $z_i(\chi) = \chi_i^{-1}(\chi)$, such that candidate i implements χ when he/she announces platform $z_i(\chi)$ where $z_i(\chi) \neq \chi$.

Lemma 2.4

If $b > \lambda c\left(\left|z_i(x_m) - x_m\right|\right)$, for both $i = L$ and R, both candidates announce $z_i(x_m)$ and implement x_m in equilibrium.
   Proof: See Appendix 2.A.3.
   This result is less interesting, and hence, I assume that at least one candidate has $b < \lambda_i c\left(\left|z_i(x_m) - x_m\right|\right)$, in what follows. Note that with asymmetric characteristics, even if one candidate, i, has $b \geq \lambda_i c\left(\left|z_i(x_m) - x_m\right|\right)$, he/she may not commit to implementing the median policy when the other candidate, j, has $b < \lambda_j c\left(\left|z_j(x_m) - x_m\right|\right)$ because i can win even if i's policy does not converge to the median policy.

2.3 Candidates with symmetric characteristics

This subsection analyzes two candidates who have symmetric cost and disutility functions and whose ideal policies are equidistant from the median policy, $x_m - x_L = x_R - x_m$.

2.3.1 Platforms

First, the policies candidates choose to implement never overlap, and they also never choose a policy that is more extreme than their own ideal policy.

Lemma 2.5

In equilibrium, the pair of platforms, $\{z_L, z_R\}$, satisfies $x_L \leq \chi_L(z_L) \leq x_m \leq \chi_R(z_R) \leq x_R$, where $x_L < x_m < x_R$.
   Proof: See Appendix 2.A.4.
   However, there is a possibility that candidates' platforms may encroach on the opponent's side of the policy space (i.e., $z_R < x_m < z_L$), which I do allow for. See Subsection 2.3.3 for more details.
   When candidate i wins, the utility of i is $-u\left(\left|\chi_i(z_i) - x_i\right|\right) - \lambda c\left(\left|z_i - \chi_i(z_i)\right|\right) + b$. When opponent j wins, the utility of i is $-u\left(\left|\chi_j(z_j) - x_i\right|\right)$. In equilibrium, these two utilities must be the same.

Proposition 2.6

Suppose $u''(d) > 0$, for any $d \geq 0$. Suppose also that two symmetric candidates choose to run. The pair of platforms, $\{z_L, z_R\}$, is an equilibrium strategy if and only if

$$-u\big(\big|\chi_i(z_i)-x_i\big|\big)-\lambda c\big(\big|z_i-\chi_i(z_i)\big|\big)+b=-u\big(\big|\chi_j(z_j)-x_i\big|\big) \qquad (2.2)$$

for i, $j = L$, R and $i \neq j$. Such an equilibrium strategy exists, and is symmetric and unique.

Proof: See Appendix 2.A.5

The main idea of the proof is as follows. When two candidates will tie, if $-u\big(\big|\chi_i(z_i)-x_i\big|\big)-\lambda c\big(\big|z_i-\chi_i(z_i)\big|\big)+b > -u\big(\big|\chi_j(z_j)-x_i\big|\big)$, each candidate prefers to be certain of winning because his/her utility will be higher than when the opponent wins. If a candidate approaches x_m, he/she is certain of winning. Therefore, the candidate will deviate in this direction. If $-u\big(\big|\chi_i(z_i)-x_i\big|\big)-\lambda c\big(\big|z_i-\chi_i(z_i)\big|\big)+b < -u\big(\big|\chi_j(z_j)-x_i\big|\big)$, the candidate would actually prefer the opponent to win. In this case, the candidate deviates away from x_m and so is certain to lose. I assume $b < \lambda c\big(\big|z_i(x_m)-x_m\big|\big)$, and hence, $\chi_i(z_i)$ and $\chi_j(z_j)$ should diverge to satisfy equation (2.2).

By contrast, if the disutility function is linear, and $x_R - x_L$ is quite small, a candidate does not mind if the opponent wins because the opponent's ideal policy is similar to his/her own ideal policy. Therefore, the candidates may prefer to stay with their ideal policies.

Corollary 2.7

Consider that $u''(d) = 0$, for all $d \geq 0$. Then, if

$$\frac{u(x_R - x_L)-b}{2} < \lambda c\big(\big|z_i(x_i)-x_i\big|\big),$$

the candidates choose $z_i = \chi_i(z_i) = x_i$ in equilibrium. Otherwise, the candidates choose $\{z_L, z_R\}$, which satisfies (2.2).

Proof: See Appendix 2.A.6.

2.3.2 Comparative statistics: cost of betrayal

Suppose the following assumption.

Assumption 1

$c'(d)/c(d)$ *strictly decreases with d, and goes to infinity as d goes to zero.*

This assumption means that the relative marginal cost decreases as $|z_i - \chi_i|$ increases. For example, if the function is monomial, this assumption holds and will be satisfied by many polynomial functions. Therefore, this assumption is quite weak.

This subsection shows the comparative statistics of the relative importance of betrayal, λ. To commit to implementing the same policy, a candidate needs to pay a larger cost of betrayal when λ decreases.

Proposition 2.8

Suppose Assumption 1 holds. Suppose also that two symmetric candidates choose to run. Then, the realized cost of betrayal, $\lambda c\big(|z_i - \chi_i(z_i)|\big)$, decreases as λ increases, given the policy to be implemented. The realized cost of betrayal goes to zero as λ goes to infinity, and the candidates' policies and platforms converge to x_m.

Proof: See Appendix 2.A.7.

Note that with complete information, voters can correctly guess the policy a candidate will implement by observing the announced platform. Thus, to win the election, the position of the policy that will be implemented is more important than the position of the platform. This is why I investigate the realized cost of betrayal given the policy to be implemented (i.e., the electoral outcome).

When λ increases, a candidate does not want to betray the platform. Therefore, $|z_i - \chi_i(z_i)|$ and $c\big(|z_i - \chi_i(z_i)|\big)$ decrease, and the decrease in $c\big(|z_i - \chi_i(z_i)|\big)$ is faster than the increase in λ. As a result, $\lambda c\big(|z_i - \chi_i(z_i)|\big)$ decreases with λ. When $\lambda c\big(|z_i - \chi_i(z_i)|\big)$ goes to zero, $b > \lambda c\big(|z_i(x_m) - x_m|\big)$ as $b > 0$. From Lemma 2.4, both candidates will implement x_m. Therefore, if λ reaches infinity, the two candidates converge to the median policy, as in the case of completely binding platforms. However, when $\lambda < \infty$, they prefer to diverge. As λ goes to zero, the policy the candidates would choose to implement converges to their respective ideal policies.[3] Therefore, completely binding and nonbinding platforms are extreme cases of partially binding platforms.

2.3.3 Position of the platforms and a probabilistic model

In my model, there is no overlap between polices to be implemented; that is, $\chi_L(z_L) \le x_m \le \chi_R(z_R)$ in equilibrium, from Lemma 2.5. However, there is a possibility that the platforms are further from the candidate's ideal policy than the median policy. In other words, platforms may encroach on the opponent's policy space; that is, $z_R < x_m < z_L$. This could happen when $u\big(|\chi_j(x_m) - x_i|\big) - u\big(|\chi_i(x_m) - x_i|\big) + b > -\lambda c\big(|x_m - \chi_i(x_m)|\big)$. If this equation holds, the candidates have an incentive to compromise more when their platforms are the same as x_m. Fortunately, this point should not be a serious problem for the following two reasons.

First, for simplification, my model assumes that candidates know every decision-relevant fact about voter preferences. If candidates are uncertain about voter preferences – that is, a probabilistic voting model is considered – the above situation does not hold in many cases. That candidates have a greater divergence of policies in a probabilistic model is well known (Calvert, 1985). Thus, the platform can enter the candidate's own side in a probabilistic voting model.

Second, the platform may encroach on the opponent's policy space. There are two main parties in Japan: the Liberal Democratic Party of Japan (LDP), which supports increased public work to sustain rural areas, and the Democratic Party of Japan (DPJ), which supports economic reforms and the reduction of government debt. In 2001, Prime Minister Junichiro Koizumi, a member of the LDP, promised to implement radical economic reforms that were also suggested by the DPJ, including a reduction in government works and debt. Thus, Koizumi and the LDP promised policies also advocated by the DPJ (Mulgan, 2002, pp. 56–57). Moreover, in the 2007 Upper House election, the LDP and Prime Minister Shinzo Abe promised continued implementation of Koizumi's economic reforms, while the DPJ promised some policies to recover and support rural areas.[4] This was a complete reversal of the original stance of the parties. Some media also indicated that Hillary Clinton seemed more conservative than John McCain in the 2008 US presidential preliminary elections. My model can explain both cases in which the platforms encroach or do not encroach on the opponent's side.

2.4 Candidates with asymmetric characteristics

2.4.1 Equilibrium

This section shows that the model of partially binding platforms can predict the winner when candidates have asymmetric characteristics (such as having asymmetric ideal policies, λ, and β). This section also shows the basic method for deriving a winner.

I denote

$$\Psi_i\left(z_i, z_j\right) \equiv -u\left(\left|\chi_i\left(z_i\right)-x_i\right|\right)-\lambda c\left(\left|z_i-\chi_i\left(z_i\right)\right|\right)-\left[-u\left(\left|\chi_j\left(z_j\right)-x_i\right|\right)\right]$$

That is, $\Psi_i\left(z_i, z_j\right)$ refers to the difference between the utility of candidate i when candidate i wins $\left(-u\left(\left|\chi_i\left(z_i\right)-x_i\right|\right)-\lambda c\left(\left|z_i-\chi_i\left(z_i\right)\right|\right)\right)$ and the utility of i when the opponent, j, wins $\left(-u\left(\left|\chi_j\left(z_j\right)-x_i\right|\right)\right)$, ignoring the fixed values, b and k. When the candidates tie, a candidate will want to make ensure its win by approaching the median policy if $\Psi_i(z_i, z_j) + b > 0$ but will want to lose if $\Psi_i(z_i, z_j) + b < 0$. The candidate with the higher $\Psi_i(z_i, z_j)$ has the greater incentive to win. Therefore, $\Psi_i(z_i, z_j)$ refers to the *degree of incentive to win*.

I also denote

$$d_i \equiv \left|\chi_i(z_i) - x_m\right| \text{ such that } \Psi_i(z_i, z_j)$$
$$+ b = 0 \text{ and } \left|\chi_i(z_i) - x_m\right| = \left|\chi_j(z_j) - x_m\right| \tag{2.3}$$

when $b < \lambda c\left(\left|z_i(x_m) - x_m\right|\right)$. That is, candidate i is indifferent between winning and losing when the opponent's policy is equidistant from x_m as the own policy, and this distance is d_i. When $b \geq \lambda c\left(\left|z_i(x_m) - x_m\right|\right)$, as a candidate has an incentive to commit to implement x_m, $\Psi_i(z_i, z_j) + b > 0$ for all symmetric pairs of $\chi_i(z_i)$ and $\chi_j(z_j)$ $\left(\Psi_i(z_i(x_m), z_j(x_m)) + b \geq 0\right)$. In this case, suppose that $d_i = 0$. From Corollary 2.3 and Proposition 2.6, (i) $\Psi_i(z_i, z_j) + b > 0$ if $\left|\chi_i(z_i) - x_m\right| = \left|\chi_j(z_j) - x_m\right| > d_i$; (ii) $\Psi_i(z_i, z_j) + b < 0$ if $\left|\chi_i(z_i) - x_m\right| = \left|\chi_j(z_j) - x_m\right| < d_i$ (and $d_i > 0$); and (iii) the value of d_i is uniquely determined. In words, if the distance between $\chi_i(z_i)$ and x_m ($\chi_j(z_j)$ and x_m) is longer than d_i, i has an incentive to win while i does not have such an incentive if this distance is shorter than d_i. Then, suppose $d_i < d_j$, that is, candidate i has an incentive to commit to a more moderate policy in the event of a tie with $\Psi_j(z_j, z_i) + b = 0$. In this situation, the following proposition shows that candidate i announces a platform such that the policy he/she will choose to implement is slightly closer to the median policy than that of j, ensuring that in equilibrium, i will win.

One technical issue is that equilibrium may not exist in a deterministic model with a continuous policy space. Suppose L wins with certainty; that is, L commits to $\left|\chi_L(z_L) - x_m\right| < \left|\chi_R(z_R) - x_m\right|$, which is a more moderate policy than that of R. In this case, L prefers to move to a more extreme policy such that L would still win against R, but the policy L would implement would be closer to his/her ideal policy. Note that such a policy exists because the policy space is continuous.

By contrast, if a discrete policy space is introduced in the above case, L may not be able to find such a policy. Suppose we have a grid of evenly spaced policies. The distance between sequential policies is $\epsilon > 0$. The other settings remain the same. Note that the purpose of introducing a discrete policy space is to ensure equilibrium, not to show new implications from a discrete case. Thus, assume that ϵ is a very small value so that the situation is almost the same as that of a continuous policy space.[5] In the following, I assume such a discrete policy space.

Proposition 2.9

Consider a case of discrete policy space. Suppose $d_i < d_j$. Then, there exists an equilibrium, and the pair of platforms $\{z_i, z_j\}$ is an equilibrium strategy if and only if

$$\psi_i(z_i, z_j) + b > 0 \text{ and } \psi_j(z_j, z_i) + b \le 0 \text{ or}$$
$$\psi_i(z_i, z_j) + b \ge 0 \text{ and } \psi_j(z_j, z_i) + b < 0, \tag{2.4}$$

where $\chi_i(z_i)$ is closer to x_m than $\chi_j(z_j)$ by ϵ; that is, $\chi_L(z_L) = x_m - (\chi_R(z_R) - x_m) + \epsilon$ if $i = L$, and $\chi_R(z_R) = x_m + (x_m - \chi_L(z_L)) - \epsilon$ if $i = R$. In equilibrium, i is certain to win, and hence, there is no equilibrium in which both candidates have the same probability of winning.

Proof: See Appendix 2.A.8.

The intuition is as follows. Suppose that candidate i has a greater incentive to approach x_m than opponent j does when they tie with $\Psi_j(z_j, z_i) + b = 0$ (i.e., $d_i < d_j$). Then, a pair of platforms exists, $\{z_i, z_j\}$, such that the policies each will choose to implement are equidistant from the median policy $\left(|\chi_i(z_i) - x_m| = |\chi_j(z_j) - x_m|\right)$ and candidate j has an incentive to lose $\left(\psi_i(z_i, z_j) + b < 0\right)$, while candidate i has an incentive to win $\left(\psi_i(z_i, z_j) + b > 0\right)$. In equilibrium, candidate i announces such a platform, while candidate j announces a slightly more extreme platform than that of i (by ϵ) and so chooses to lose. Figure 2.1 shows the policies to be implemented, given such platforms.

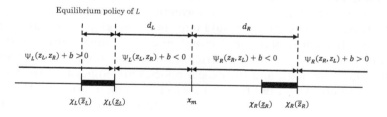

Figure 2.1 Candidates Having Asymmetric Characteristics.

Suppose $d_L < d_R$. Then, candidate L has an incentive to win ($\psi_L(z_L, z_R) + b > 0$), while candidate R does not have it ($\psi_R(z_R, z_L) + b < 0$) when both candidates' *symmetric* policies are within the bold area. Candidate L announces a platform such that his/her policy is within the bold area, and R loses.

Note that in this equilibrium, i wins, and j loses with certainty. There does not exist any equilibrium where a candidate's probability of winning is less than 1 and more than 0. Note also that as the policy implemented by the winner is only slightly (ϵ) closer to the median policy than that of the loser, vote shares between the two candidates are very close to 50%.

Equilibrium satisfies $\Psi_i(z_i, z_j) + b > 0$ and $\Psi_j(z_j, z_i) + b \leq 0$, or $\Psi_i(z_i, z_j) + b \geq 0$ and $\Psi_j(z_j, z_i) + b < 0$. As a result, multiple equilibria exist. Denote $\overline{z}_i$ as the most extreme platform of i, and $\underline{z}_i$ as the most moderate platform of i among all possible equilibrium platforms. More precisely, $\overline{z}_i$ satisfies $\Psi_j(\overline{z}_j, \overline{z}_i) + b = 0$ where $|\chi_i(\overline{z}_i) - x_m| = |\chi_j(\overline{z}_j) - x_m|$, and $\underline{z}_i$ satisfies $\Psi_i(\underline{z}_i, \underline{z}_j) + b = 0$ where $|\chi_i(\underline{z}_i) - x_m| = |\chi_j(\underline{z}_j) - x_m|$ if $b < \lambda_i c(|z_i(x_m) - x_m|)$. If $b \geq \lambda_i c(|z_i(x_m) - x_m|)$, $\underline{z}_i = z_i(x_m)$, which is the platform committing to implement the median policy. Any platform between $\overline{z}_i$ and $\underline{z}_i$ can be an equilibrium strategy of the winner i. Figure 2.1 also shows the positions of $\chi_i(\overline{z}_i)$ and $\chi_i(\underline{z}_i)$.

2.4.2 Winner of an asymmetric election

From Proposition 2.9, if candidate i has a greater incentive to approach x_m (i.e., $\psi_i(z_i, z_j)$ is greater than $\psi_j(z_j, z_i)$ in the event of a tie with $\psi_j(z_j, z_i) + b = 0$), then, in equilibrium, candidate i always wins. This implies that to find the winner of an asymmetric election, it is sufficient to compare candidates' degrees of incentive to win. This can be given as follows.

In order to prove that i wins against j, assume that the policies each candidate would implement are initially fixed at symmetric positions (i.e., $|\chi_i(z_i) - x_m| = |\chi_j(z_j) - x_m|$). This implies that the electoral outcome (a tie) is fixed by fixing the policies to be implemented. Note that because voters have complete information, they can correctly guess the policy each candidate would implement, making the positions of these policies critical to the electoral outcome. Suppose also that two candidates are initially symmetric (i.e., they have symmetric cost and disutility functions, and their ideal policies are equidistant from the median policy) and indifferent between winning and losing, that is, $\Psi_i(z_i, z_j) + b = \Psi_j(z_j, z_i) + b = 0$. Then, differentiate $\Psi_j(z_j, z_i)$ by the parameter of a candidate's characteristic (such as x_j, λ_j, or β_j). Now, suppose j's parameter value is higher than that of i. If $\Psi_j(z_j, z_i)$ decreases with this parameter value, it means that $\Psi_j(z_j, z_i)$ is lower than $\Psi_i(z_i, z_j)$ in a tie with $\Psi_j(z_j, z_i) + b = 0$. Hence, i is certain to win, according to Proposition 2.9.

In the following subsections, I use the above method to show the asymmetric electoral outcomes for asymmetric ideal policies, asymmetric costs of betrayal, and asymmetric policy motivations. Although I only consider these basic characteristics, this model could be used to derive more implications by adding other characteristics (e.g., competence and valence) or other players (e.g., special interest groups and media).

2.4.3 Asymmetric ideal policies

Assume that $x_R - x_m \neq x_m - x_L$; that is, the candidate's ideal policy is asymmetric. The cost and disutility functions are the same for both candidates. Suppose also the following assumption.

Assumption 2

$u''(d) / u'(d)$ *is non-increasing in* d.

This assumption means that the Arrow-Pratt measure of absolute risk aversion is non-increasing in $|\chi_i(z_i) - x_i|$. If the function is monomial, this assumption holds and will be satisfied by many polynomial functions.

Corollary 2.10

Suppose Assumptions 1 and 2, and that two candidates run. Furthermore, suppose that candidate i is more moderate (i.e., $|x_i - x_m| < |x_j - x_m|$), but that the candidates are symmetric in all other respects. Then, in equilibrium, we have the following: (i) when $u''(d) > 0$, for any $d > 0$, candidate i wins with certainty and the expected utility from winning is higher than the expected utility from losing; and (ii) when $u''(d) = 0$, for all $d \geq 0$, the result is either a tie or candidate i wins with certainty.

Proof: See Appendix 2.A.9.

A more moderate candidate, whose ideal policy is closer to the median policy, will not severely betray his/her platform after an election. On the other hand, in order to implement the same policy, a more extreme candidate will pay a higher cost of betrayal because he/she will betray the platform more severely. Hence, his/her degree of incentive to win decreases to avoid paying such a high cost of betrayal. As a result, the more moderate candidate wins.

When the candidates' utility functions are linear, they tie in most cases. When a candidate has a linear utility function, the policy he/she will implement is not affected by the ideal policy, x_i. Therefore, the situation is the same for both candidates and they have the same probability of winning. However, if the moderate candidate's ideal policy is very close to x_m, the moderate candidate does not have an incentive to approach x_m from his/her ideal policy, and the extreme candidate does not have an incentive to win even though the moderate candidate announces x_i. Thus, the moderate candidate announces his/her ideal policy as the platform and then implements it after he/she wins. This case is similar to Corollary 2.7, but the moderate candidate wins with certainty, as his/her ideal policy is closer to the median policy.

2.4.4 Asymmetric costs of betrayal

Assume that λ is not the same for both candidates, and Candidate i has λ_i. However, their ideal policies and disutility functions are symmetric.

Corollary 2.11

Suppose Assumption 1, and that two candidates run. Suppose also that candidate i has a higher relative importance of betrayal (i.e., $\lambda_i > \lambda_j$), but that the candidates are symmetric in all other respects. Then, in equilibrium, candidate i wins with certainty. The expected utility from winning is higher than, or the same as, the expected utility from losing.

Proof: See Appendix 2.A.10.

When a candidate has a lower λ_i, he/she will betray the platform more severely, and hence, the realized cost of betrayal is higher, as shown in Proposition 2.8. Therefore, such a candidate has a lower degree of incentive to win because he/she wishes to avoid paying the high cost of betrayal. As a result, the candidate with the higher λ_i wins.

2.4.5 Asymmetric political motivations

Suppose that the level of political motivation, β, differs from 1, and Candidate i has β_i. Furthermore, assume that β_i is not the same for both candidates. However, their ideal policies and cost functions are symmetric. That is, the utility following a win is $-\beta_i u\big(|\chi_i(z_i) - x_i|\big) - \lambda c\big(|z_i - \chi_i(z_i)|\big) + b$ and the utility when the opponent wins is $-\beta_i u\big(|\chi_j(z_j) - x_i|\big)$. Thus, the degree of incentive to win is

$$\Psi_i(z_i, z_j) \equiv -\beta_i u\big(|\chi_i(z_i) - x_i|\big) - \lambda c\big(|z_i - \chi_i(z_i)|\big) + \beta_i u\big(|\chi_j(z_j) - x_i|\big).$$

Corollary 2.12

Suppose Assumption 1, and that two candidates run. Suppose also that candidate i is less policy motivated (i.e., $\beta_i < \beta_j$), but that the candidates are symmetric in all other respects. Then, in equilibrium, candidate i wins with certainty. The expected utility from winning is higher than the expected utility from losing.

Proof: See Appendix 2.A.11.

A less policy-motivated candidate is less concerned about policy and does not betray the platform so severely, and hence has a lower cost of betrayal and a higher degree of incentive to win. By contrast, a more policy-motivated candidate will betray the platform more severely, which induces a higher cost of betrayal. As a result, a less policy-motivated candidate wins the election.

2.4.6 *Functional example*

This subsection shows a functional example as an overview of the implications described so far. Suppose a linear disutility function, $\beta_i u(|\chi - x_i|) = \beta_i |\chi - x_i|$, and a quadratic cost function, $\lambda_i c(|z_i - \chi_i(z_i)|) = \lambda_i (z_i - \chi_i(z_i))^2$.

From (2.1), the policies to be implemented are

$$\chi_L(z_L) = z_L - \frac{\beta_L}{2\lambda_L},$$

$$\chi_R(z_R) = z_R + \frac{\beta_R}{2\lambda_R},$$

assuming $z_L - \beta_L/(2\lambda_L) > x_L$ and $z_R + \beta_R/(2\lambda_R) < x_R$ (Corollary 2.2). The cost of betrayal is $\beta_i^2/(4\lambda_i)$, which decreases with λ_i (Proposition 2.8).

Then, the degrees of incentive to win are

$$\psi_R(z_R, z_L) = \beta_R(\chi_R(z_R) - \chi_L(z_L)) - \frac{\beta_R^2}{4\lambda_R},$$

$$\psi_L(z_L, z_R) = \beta_L(\chi_R(z_R) - \chi_L(z_L)) - \frac{\beta_L^2}{4\lambda_L}.$$

First, when b is sufficiently high that $b > \beta_i^2/(4\lambda_i)$ for both $i = L$ and R, both candidates have an incentive to win, even if $\chi_R(z_R) = \chi_L(z_L) = x_m$. Thus, both candidates commit to implementing the median policy, and they tie (Lemma 2.4).

Now, suppose that at least one candidate has $b < \beta_i^2 / (4\lambda_i)$, and the ideal policies of L and R are symmetric. For a symmetric pair of $\chi_L(z_L)$ and $\chi_R(z_R)$, i is indifferent between winning and losing if $\psi_i(z_i, z_j) + b = 0$, that is,

$$\chi_R(z_R) - \chi_L(z_L) = \frac{\beta_i}{4\lambda_i} - \frac{b}{\beta_i}$$

which is $2d_i$ according to (2.3) when $\beta_i / (4\lambda_i) - b / \beta_i \geq 0$. If $\max\{0, \beta_L / (4\lambda_L) - b / \beta_L\} < \beta_R / (4\lambda_R) - b / \beta_R$ ($d_L < d_R$), there exists a symmetric pair of $\chi_L(z_L)$ and $\chi_R(z_R)$ such that

$$\max\left\{0, \frac{\beta_L}{4\lambda_L} - \frac{b}{\beta_L}\right\} \leq \chi_R(z_R) - \chi_L(z_L) < \frac{\beta_R}{4\lambda_R} - \frac{b}{\beta_R},$$

or

$$\max\left\{0, \frac{\beta_L}{4\lambda_L} - \frac{b}{\beta_L}\right\} < \chi_R(z_R) - \chi_L(z_L) \leq \frac{\beta_R}{4\lambda_R} - \frac{b}{\beta_R}.$$

Note that as $b < \beta_R^2 / (4\lambda_R)$, $\beta_R / (4\lambda_R) - b / \beta_R > 0$. In equilibrium, L commits to implementing $\chi_L(z_L)$, which satisfies the above condition. Then, R does not have an incentive to commit to $\chi_R(z_R)$ such that $\chi_R(z_R) - x_m = x_m - \chi_L(z_L)$, and thus, L wins with certainty. More precisely, in equilibrium, L commits to implementing $\chi_L(z_L)$, which satisfies

$$x_m - \left(\frac{\beta_R}{8\lambda_R} - \frac{b}{2\beta_R}\right) \leq \chi_L(z_L) \leq \min\left\{x_m, x_m - \left(\frac{\beta_L}{8\lambda_L} - \frac{b}{2\beta_L}\right)\right\}$$

and R commits to implementing $x_m + (x_m - \chi_L(z_L)) + \epsilon$ (Proposition 2.9). In this case, L announces $z_L \in [\underline{z}_L, \overline{z}_L]$, such that

$$\overline{z}_L = x_m - \left(\frac{\beta_R}{8\lambda_R} - \frac{b}{2\beta_R}\right) + \frac{\beta_L}{2\lambda_L},$$

$$\underline{z}_L = \min\left\{x_m + \frac{\beta_L}{2\lambda_L}, x_m - \left(\frac{\beta_L}{8\lambda_L} - \frac{b}{2\beta_L}\right) + \frac{\beta_L}{2\lambda_L}\right\}$$

The value of $\beta_i / (4\lambda_i) - b / \beta_i$ decreases with λ_i and increases with β_i. Thus, if λ_L is higher than λ_R (with $\beta_L = \beta_R$), L wins (Corollary 2.11). If β_L is lower than β_R (with $\lambda_L = \lambda_R$), L also wins with certainty (Corollary 2.12).

Suppose that $x_R - x_m > x_m - x_L$; that is, R is a more extreme candidate than L, and $\lambda_L = \lambda_R$ and $\beta_L = \beta_R$. In this case, both candidates tie in equilibrium if R still has an incentive to win $(\psi_R(z_R, z_L) + b \geq 0)$ when L chooses $x_L = z_L = \chi_L(z_L)$. In this equilibrium, candidates commit to implementing

$$\chi_L(z_L) = x_m - \left(\frac{\beta}{8\lambda} - \frac{b}{2\beta}\right),$$

$$\chi_R(z_R) = x_m + \left(\frac{\beta}{8\lambda} - \frac{b}{2\beta}\right),$$

which satisfy (2.2). By contrast, if R does not have an incentive to win against L when L chooses $x_L = z_L = \chi_L(z_L)$; that is,

$$2\beta(x_m - x_L) - \frac{\beta^2}{4\lambda} + b < 0$$

then L chooses $x_L = z_L = \chi_L(z_L)$, and R commits to implementing a more extreme policy than x_L. In this case, L wins with certainty in equilibrium (Corollary 2.10(ii)).

2.4.7 Applications

The model of partially binding platforms can be applied to some other topics as follows.

2.4.7.1 Maturity of democracy

The value of λ is decided by many factors. For example, when the freedom of the press is curtailed, λ is low because the media will not report politicians' betrayals. When a large special interest group supports politicians, the politicians are assured of a large number of votes in an election, and the probability of their losing the next election is quite low. Therefore, in this case, the candidate is less concerned about the cost of betrayal. If a party or the parliament is lacking in power, λ is low because these institutions are less able to enforce discipline.

In other words, λ can be interpreted as the level of a democracy's maturity. Some political scientists and economists indicate that politicians in mature democracies have a greater ability to make binding platforms. For example, in immature democracies, politicians have strong relationships with specific groups of voters.[6] If the democracy is mature, it supports freedom of the press and government transparency. In addition, strong parties monitor politicians, who therefore do not betray their platforms as often or as easily.[7] Thus, the value of λ is higher in mature democracies and lower in immature democracies.

In fact, using cross-country data, Keefer (2007) shows the differences between younger and older democracies, which tend to arise from the inability of younger democracies to offer credible platforms to voters.

According to Proposition 2.8, when the maturity of a democracy increases, the policies to be implemented converge to x_m, and politicians do not often renege on their platforms. In an immature democracy, the divergence in policies to be implemented is large, and politicians tend to betray their platforms quite severely.

2.4.7.2 Endogenous cost of betrayal

The candidate may make decisions that affect the value of λ. For example, sometimes he/she decides to influence the media. If the candidate is able to control the media, the cost of betrayal, λ, decreases, and he/she can betray the platform more easily. This seems favorable to candidates, although they usually support freedom of the press, even when the media criticize them.

There is another case. In Japan, since 2003, the Democratic Party of Japan has issued manifestos. In a manifesto, the party records its platform, allowing voters and the media to compare it to the policy implemented after the election. Before 2003, candidates and parties revealed their platforms in speeches, campaign posters, and discussions with the media, but there were no official written records of their platforms. Thus, after 2003, it became easier to check whether the governing party betrayed its platforms. For parties, the publication of a manifesto increases the cost of betrayal, which seems detrimental to their interests. However, other parties also began issuing manifestos from 2003 onward (Kanai, 2003).

One reason is that a higher λ means a higher probability of winning. Moreover, as the expected utility from winning is not lower than the expected utility from losing (Corollary 2.11), if candidates can change λ, they will choose a value that is as high as possible in equilibrium. Sometimes, politicians prefer to use explicit and impressive words, promising, for example, to "end welfare as we know it." Such words are easy to remember, and hence, increase the value of λ.

2.4.7.3 Seniority of candidates

Older (or more senior) politicians may have a lower value of λ. They tend to be less concerned about the next election or their party's discipline because they may retire before the next election. In such a case, the value of λ could be asymmetric. According to my model with asymmetric candidates, if a candidate is older, he/she will betray the platform more severely, and hence, the probability of winning decreases. This is one type of the "last-term problem."[8]

However, at the same time, political motivation may also differ depending on seniority. Younger politicians may care more about policy (and have a higher β) than older politicians, as many of these policies are likely to impact the younger candidate's future political career.

Suppose R is more policy motivated but has a higher relative importance of betrayal, that is, $\beta_L < \beta_R$ and $\lambda_L < \lambda_R$ (and $x_R - x_m = x_m - x_l$). According to my interpretation, R is younger (or less senior) than L. From the numerical example, if

$$ b\left(\frac{1}{\beta_L} - \frac{1}{\beta_R}\right) > \frac{1}{4}\left(\frac{\beta_L}{\lambda_L} - \frac{\beta_R}{\lambda_R}\right) \tag{2.5} $$

L wins with certainty. As $\beta_L < \beta_R$, the left-hand side of (2.5) is positive. If $\lambda_L / \lambda_R \geq \beta_L / \beta_R$, L still wins with certainty as the right-hand side of (2.5) is non-positive. However, it is not obvious if $\lambda_L / \lambda_R < \beta_L / \beta_R$. In this case, if b is sufficiently high, L wins with certainty, but if b is sufficiently low, R wins with certainty because, if b is sufficiently large, the difference in λ is not so critical. One possible interpretation of b is related to politicians' wages. Thus, the above implication suggests that a higher wage induces older (or more senior) politicians to win, while a lower wage induces younger (or less senior) politicians to win. For instance, by using the data on Brazil's municipal government, Ferraz and Finan (2011) show that a higher pay induces senior politicians to win.

Similarly, the model of partially binding platforms can show and analyze asymmetric electoral outcomes for candidates with different characteristics and is applicable in cases other than the above topics as well.

2.5 Decisions to run

This section analyzes candidates' decisions to run. There are two potential candidates in the district, and they decide whether to run for office. There are two possible cases: (1) the two potential candidates are symmetric and (2) they are asymmetric (about ideal policies, λ, and β).

In a two-candidate competition, when one candidate deviates by withdrawing, the remaining candidate who runs can announce his/her own ideal policy as his/her platform and will implement it after an election because he/she no longer has a rival.

2.5.1 Symmetric two-candidate equilibrium

Suppose that two potential candidates are symmetric. As in Section 2.3, suppose $\beta_i = \beta_{j_*} = 1$ and $\lambda_i = \lambda_j = \lambda$. Suppose also $u''(d) > 0$ for any $d > 0$. Denote z_i^* and z_j^* as the unique equilibrium pair of platforms when two symmetric candidates run; that is, equation (2.2) holds. For candidate i, the utility when he/she runs is $(1/2)\left[-u\left(\left|\chi_i\left(z_i^*\right)-x_i\right|\right)-\lambda c\left(\left|z_i^*-\chi_i\left(z_i^*\right)\right|\right)-u\left(\left|\chi_j\left(z_j^*\right)-x_i\right|\right)+b\right]-k$. The utility of i when he/she does not run is $-u\left(\left|x_i-x_j\right|\right)$. Because condition (2.2) holds, $(1/2)\left[-u\left(\left|\chi_i\left(z_i^*\right)-x_i\right|\right)-\lambda c\left(\left|z_i^*-\chi_i\left(z_i^*\right)\right|\right)-u\left(\left|\chi_j\left(z_j^*\right)-x_i\right|\right)+b\right]=-u\left(\left|\chi_j\left(z_j^*\right)-x_i\right|\right)$. Therefore, both candidates do not deviate by withdrawing if

$$k \leq u\left(\left|x_i-x_j\right|\right)-u\left(\left|\chi_j\left(z_j^*\right)-x_i\right|\right) \tag{2.6}$$

for i, $j = L$, R and $i \neq j$. If (2.6) does not hold, an equilibrium where only one candidate runs exists. In such an equilibrium, one candidate announces his/her ideal policy as a platform and implements it after an election. Note that as the payoff for all players is $-\infty$ if no one runs, this candidate never deviates by not running. Therefore, if k is sufficiently low and/or $\left|x_i-x_j\right|$ is sufficiently large, a symmetric two-candidate equilibrium exists.

Corollary 2.13

Suppose $u''(d) > 0$ for any $d > 0$. A symmetric two-candidate equilibrium exists if two potential candidates satisfy equation (2.6). Otherwise, one potential candidate runs and wins.

2.5.2 Asymmetric two-candidate equilibrium

In general, potential candidates may not be symmetric, and hence, I suppose two asymmetric potential candidates. Suppose candidate j is the loser (with a more extreme ideal policy, lower relative importance of betrayal, or more policy motivation) and candidate i is the winner. Furthermore, suppose i announces $\bar{z}_i$, as introduced in Subsection 2.4.1. For the loser, j, the utility when he/she runs is $-\beta_j u\left(\left|\chi_i\left(\bar{z}_i\right)-x_j\right|\right)-k$. The utility of j when he/she does not run is $-\beta_j u\left(\left|x_i-x_j\right|\right)$. It is always $-u\left(\left|\chi_i\left(\overline{z}_i\right)-x_j\right|\right) \geq -u\left(\left|x_i - x_j\right|\right)$. Thus, the loser j does not deviate by withdrawing, if

$$k \leq \beta_j u\left(\left|x_i - x_j\right|\right) - \beta_j u\left(\left|\chi_i\left(\overline{z}_i\right) - x_j\right|\right) \tag{2.7}$$

For the winner i, the utility when he/she runs is $-\beta_i u\left(\left|\chi_i\left(\overline{z}_i\right) - x_i\right|\right) - \lambda_i c\left(\left|\overline{z}_i - \chi_i\left(\overline{z}_i\right)\right|\right) + b - k$. The utility of i when he/she does not run is $-\beta_i u\left(\left|x_i - x_j\right|\right)$. Thus, the winner i does not deviate by not running, if

$$k \leq \beta_i u\left(\left|x_i - x_j\right|\right) - \beta_i u\left(\left|\chi_i\left(\overline{z}_i\right) - x_i\right|\right) - \lambda_i c\left(\left|\overline{z}_i - \chi_i\left(\overline{z}_i\right)\right|\right) + b \tag{2.8}$$

As I have shown, in any case (Corollaries 2.10–2.12), the winner's expected utility is higher than or the same as the loser's expected utility, and hence, $-\beta_i u\left(\left|\chi_i\left(\overline{z}_i\right) - x_i\right|\right) - \lambda_i c\left(\left|\overline{z}_i - \chi_i\left(\overline{z}_i\right)\right|\right) + b \geq -\beta_j u\left(\left|\chi_i\left(\overline{z}_i\right) - x_j\right|\right)$. However, if candidates have asymmetric policy motivations, $\beta_i < \beta_j$, which means $\beta_i u\left(\left|x_i - x_j\right|\right) < \beta_j u\left(\left|x_i - x_j\right|\right)$. When $\beta_j - \beta_i$ is sufficiently small and b is high, (2.8) holds if (2.7) holds. This implies that when the loser j does not deviate, the winner i also does not deviate. When $\beta_j - \beta_i$ is sufficiently large and b is small, (2.7) holds if (2.8) holds. This means that when the winner i does not deviate, the loser j also does not deviate. Note that a sufficiently large $\beta_j - \beta_i$ means that β_i is small. This means that candidate i cannot get sufficient benefits even if he/she is certain to run and win, and hence (2.8) becomes a critical condition in this case. In either case, if the cost of running is sufficiently small, then both asymmetric candidates do not have an incentive to deviate by withdrawing.

If (2.7) or (2.8) is not satisfied, an equilibrium exists in which only i or j runs. If (2.7) is satisfied with inequality, but (2.8) is not satisfied, only candidate j runs. Similarly, if (2.8) is satisfied with inequality, but (2.7) is not satisfied, only candidate i runs.[9] If neither inequality is satisfied, either of the two potential candidates runs.[10] In either case, one candidate announces his/her ideal policy and implements it. Therefore, if k is sufficiently low and/or $\left|x_i - x_j\right|$ is sufficiently large, an asymmetric, two-candidate equilibrium exists.

Note that candidate i's platform $\overline{z}_i$ is the most extreme platform among all possible equilibrium platforms. Thus, if $\overline{z}_i$ satisfies (2.7) and (2.8), other possible equilibrium platforms also satisfy them.

Proposition 2.14

An asymmetric two-candidate equilibrium exists if two potential candidates satisfy equations (2.7) and (2.8).

In such an equilibrium, the loser j runs in order to induce the winner i to approach j's ideal policy even though j loses the election. If j deviates by not running, i will implement his/her ideal policy x_i. On the other hand, i will approach the median policy (and hence, j's ideal policy) more closely if j runs. Therefore, j runs to induce i to approach x_j, even though it is certain j will lose.

In the models of non-binding platforms, the winner will implement his/her ideal policy after an election, which means that the loser's decision to run does not affect the winner's policy. Thus, the loser does not have any reason to run. In the models of completely binding platforms, as both candidates have an equal probability (50%) of winning, an explicit loser does not exist. Thus, the setting of a partially binding platform is important to derive such strategic behaviors.

2.5.3 *Application: the Social Democratic Party of Japan*

In Japan, the LDP was established in 1955, and it was the party in power right from 1955 to 1993. Sartori (1976) uses it as one example of the predominant-party system: the major party consistently forms the government by winning a majority, even though there is a completely competitive electoral system.

The largest opposition party in Japan, the Social Democratic Party of Japan (SDPJ), was also established in 1955. However, in 1960, the conservative faction left the SDPJ and established a new third party. After this event, the SDPJ was unable to win a majority, marking the beginning of the LDP's predominant-party system. Except in the 1958 election, the SDPJ ran candidates for less than 50% of the legislative seats in all elections, with the number of SDPJ candidates decreasing over time. The SDPJ's policy position shifted further to the left (socialism) after the conservative faction left, and the SDPJ remained adamant about its stance on policy issues, not wishing to compromise its core beliefs to appeal to the voter base. The SDPJ was not able to form the central government until 1994, but it was able to change or reject bills proposed by the LDP through negotiations or by using their limited veto power, such as boycotts (Ihori and Doi, 1998).[11]

The above strategies and incentives of the SDPJ match the model in this section: The SDPJ gave up trying to gain the majority of seats (i.e., accepted defeat) even though it even though it could have pacified voters by adopting a softer stance (i.e., approached the median policy). This result may have occurred because the SDPJ might have been content with only modifying the LDP's policies to bring them closer to SDPJ's ideal policies, instead of trying to be the main party of Japan.

2.6 Summary

This chapter examined the effects of partially binding platforms in elections. Although a candidate can choose any policy that differs from the promised policy before an election, it is costly to do so. I extend the standard political-competition model introduced by Downs (1957) by considering that candidates choose a platform and a policy separately, and by also adding the cost to renege on the platform. The model of partially binding platforms shows the following two notable implications, which cannot be analyzed by previous formal models. First, when candidates have different characteristics, one candidate has a higher probability of winning. That is, asymmetric electoral outcomes can be analyzed. Second, even if a candidate knows that he/she will lose an election, this loser runs to induce the opponent to approach the loser's ideal policy. That is, it can describe the loser's incentive more clearly.

2.A Appendix: proofs

2.A.1 Corollary 2.2

With $u''(d) = 0$, for all $d \geq 0$, condition (2.1) is

$$\lambda = \frac{\bar{u}}{c'\left(\left|z_i - \chi_i(z_i)\right|\right)},$$

which does not depend on the position of x_i. If $\lambda < \bar{u} / c'\left(\left|z_i - x_i\right|\right)$, $\chi_i(z_i)$, which satisfies (2.1), is further away from z_i than x_i. However, if a candidate chooses x_i rather than this $\chi_i(z_i)$, the disutility is minimized at $u(0) = 0$, and the cost of betrayal also decreases. Thus, the winner implements x_i if $\lambda < \bar{u} / c'\left(\left|z_i - x_i\right|\right)$.

∎

2.A.2 Corollary 2.3

Suppose Candidate L, without loss of generality. Note that an increase in z_L means that z_L moves further from the ideal policy, x_L. Rewrite

(2.1) as $\lambda c'(z_L - \chi_L(z_L)) = u'(\chi_L(z_L) - x_L)$ and differentiate it by z_L. Then, it becomes

$$\lambda c''(z_L - \chi_L(z_L))\left(1 - \frac{\partial \chi_L(z_L)}{\partial z_L}\right) = u''(\chi_L(z_L) - x_L)\frac{\partial \chi_L(z_L)}{\partial z_L}$$

$$\Rightarrow \frac{\partial \chi_L(z_L)}{\partial z_L} = \frac{\lambda c''(z_L - \chi_L(z_L))}{u''(\chi_L(z_L) - x_L) + \lambda c''(z_L - \chi_L(z_L))} \in (0,1).$$

Differentiate $u(\chi_L(z_L) - x_L)$ by z_L. Then,

$$\frac{\partial u(\chi_L(z_L) - x_L)}{\partial z_L} = u'(\chi_L(z_L) - x_L)\frac{\partial \chi_L(z_L)}{\partial z_L} > 0.$$

Differentiate $\lambda c(z_L - \chi_L(z_L))$ by z_L. Then,

$$\frac{\partial \lambda c(z_L - \chi_L(z_L))}{\partial z_L} = \lambda c'(z_L - \chi_L(z_L))\left(1 - \frac{\partial \chi_L(z_L)}{\partial z_L}\right) > 0.$$

∎

2.A.3 Lemma 2.4

Existence

Suppose that both candidates commit to implementing the median policy $(\chi_i(z_i) = \chi_j(z_j) = x_n)$. Then, the expected utility for candidate i is $-u(|x_m - x_i|) + (1/2)(-\lambda c(|z_i(x_m) - x_m|) + b)$. If i deviates to lose by committing to a more extreme policy, i's expected utility is $-u(|x_m - x_i|)$. Thus, if $b > \lambda c(|z_i(x_m) - x_m|)$, candidate i does not deviate.

Uniqueness

First, suppose that $\chi_i(z_i) \neq x_n$, $\chi_j(z_j) \neq x_n$, and $|\chi_i(z_i) - x_m| > |\chi_j(z_j) - x_m|$: thus, i loses. The expected utility of i is $-u(|\chi_j(z_j) - x_i|)$. If i deviates to commit to a policy, χ_i', which is slightly closer to x_m than $\chi_j(z_j)$, and which is also closer to x_i than x_m, the expected utility is $-u(|\chi_i' - x_i|) - \lambda c(|z_i(\chi_i') - \chi_i'|) + b$ since i will win for sure. Thus, if $u(|\chi_j(z_j) - x_i|) - u(|\chi_i' - x_i|) - \lambda c(|z_i(\chi_i') - \chi_i'|) + b > 0$, candidate i deviates. First, $-\lambda c(|z_i(\chi_i') - \chi_i'|) + b > 0$ since $\lambda c(|z_i(\chi_i') - \chi_i'|) < \lambda c(|z_i(x_m) - x_m|)$ from Corollary 2.3, and $b > \lambda c(|z_i(x_m) - x_m|)$. Second, if $\chi_j(z_j)$ is further away from x_i than x_m, $u(|\chi_j(z_j) - x_i|) - u(|\chi_i' - x_i|) > 0$. If $\chi_j(z_j)$ is closer to x_i than x_m, $u(|\chi_j(z_j) - x_i|) - u(|\chi_i' - x_i|)$ is only slightly lower than but almost the same as zero. Thus, $u(|\chi_j(z_j) - x_i|) - u(|\chi_i' - x_i|) - \lambda c(|z_i(\chi_i') - \chi_i'|) + b > 0$.

Second, suppose that $\chi_i(z_i) \neq x_m$, $\chi_j(z_j) \neq x_m$, and $|\chi_i(z_i) - x_m| = |\chi_j(z_j) - x_m|$: thus, there is a tie. The expected utility of i is $(1/2)\big(-u(|\chi_i(z_i) - x_i|) - \lambda c(|z_i - \chi_i(z_i)|) + b - u(|\chi_j(z_j) - x_i|)\big)$. If i deviates to commit to a policy, χ_i', which is slightly closer to x_m than $\chi_i(z_i)$, and which is also closer to x_i than x_m, the expected utility is $-u(|\chi_i' - x_i|) - \lambda c(|z_i(\chi_i') - \chi_i'|) + b$ since i will win for sure. From the same reason as above, $-u(|\chi_i' - x_i|) - \lambda c(|z_i(\chi_i') - \chi_i'|) + b > -u(|\chi_j(z_j) - x_i|)$. If $\chi_i(z_i)$ is further away from x_i than x_m, $-u(|\chi_i(z_i) - x_i|) - \lambda c(|z_i - \chi_i(z_i)|) < -u(|\chi_i' - x_i|) - \lambda c(|z_i(\chi_i') - \chi_i'|)$ If $\chi_i(z_i)$ is closer to x_i than x_m, $-u(|\chi_i' - x_i|) - \lambda c(|z_i(\chi_i') - \chi_i'|)$ is only slightly lower than but almost the same as $-u(|\chi_i(z_i) - x_i|) - \lambda c(|z_i - \chi_i(z_i)|)$. Thus, $(1/2)(-u(|\chi_i(z_i) - x_i|) - \lambda c(|z_i - \chi_i(z_i)|) + b - u(|\chi_j(z_j) - x_i|)$ is lower than $-u(|\chi_i' - x_i|) - \lambda c(|z_i(\chi_i') - \chi_i'|) + b$, and as such, i deviates in this way.

Third, suppose that $\chi_i(z_i) \neq x_m$ and $\chi_j(z_j) = x_m$: thus, i loses. The expected utility of i is $-u(|x_m - x_i|)$. If i deviates to commit a policy x_m, the expected utility is $-u(|x_m - x_i|) + (1/2)\big(-\lambda c(|z_i(x_m) - x_m|) + b\big)$ since they tie. Because $b > \lambda c(|z_i(x_m) - x_m|)$, i deviates in this way. Thus, when at least one candidate does not choose, here is no equilibrium in the case of $b > \lambda c(|z_i(x_m) - x_m|)$.

■

2.A.4 Lemma 2.5

First, suppose that L's policy is more extreme than his/her ideal policy, $\chi_L(z_L) < x_L$. That is, $\chi_L(z_L) < x_L < x_m < x_R < x_m + (x_m - \chi_L(z_L))$. There are five possible positions of $\chi_R(z_R)$: (1) $\chi_R(z_R) \leq \chi_L(z_L)$; (2) $x_m + (x_m - \chi_L(z_L)) \leq \chi_R(z_R)$; (3) $\chi_L(z_L) < \chi_R(z_R) < x_R$; (4) $x_R < \chi_R(z_R) < x_m + (x_m - \chi_L(z_L))$; and (5) $\chi_R(z_R) = x_R$.

- In cases (1) and (2), L wins with certainty or has a 50% probability of winning. Both candidates have an incentive to deviate to choose $z_i = x_i = \chi_i(z_i)$ and so definitely win with the maximized expected utility from winning (i.e., b).
- In cases (3) and (4), R wins with certainty. Here, R has an incentive to deviate to choose $z_R = x_R = \chi_R(z_R)$ but still definitely wins with the maximized expected utility from winning.

- In case (5), R wins with certainty. Here, L has an incentive to deviate to choose $z_L = x_L = \chi_L(z_L)$ and so has a 50% chance of winning with the maximized expected utility from winning.

For the same reasons, if $x_R < \chi_R(z_R)$, there is no equilibrium.

Next, suppose that L's policy encroaches on R's side of the policy space, $x_m < \chi_L(z_L)$. Here, there are three possible positions of $\chi_L(z_L)$: (A) $x_m - (\chi_L(z_L) - x_m) < x_L < x_m < x_R < \chi_L(z_L)$; (B) $x_m - (\chi_L(z_L) - x_m) = x_L < x_m < x_R = \chi_L(z_L)$; and (C) $x_L < x_m - (\chi_L(z_L) - x_m) < x_m < \chi_L(z_L) < x_R$. In each case, there are five possible positions of $\chi_R(z_R)$: (1) $\chi_R(z_R) < x_m - (\chi_L(z_L) - x_m)$; (2) $\chi_L(z_L) < \chi_R(z_R)$; (3) $\chi_R(z_R) = x_m - (\chi_L(z_L) - x_m)$; (4) $\chi_L(z_L) = \chi_R(z_R)$; and (5) $x_m - (\chi_L(z_L) - x_m) < \chi_R(z_R) < \chi_L(z_L)$.

- Suppose (A):
  - In cases (1) and (2), L wins with certainty. Here, L has an incentive to deviate to choose $z_L = x_L = \chi_L(z_L)$ and still definitely wins with the maximized expected utility from winning.
  - In cases (3) and (4), they tie. Both candidates have an incentive to deviate to choose $z_i = x_i = \chi_i(z_i)$ and definitely win with the maximized expected utility from winning.
  - In case (5), R wins with certainty. If $\chi_R(z_R) \neq x_R$, R has an incentive to deviate to choose $z_R = x_R = \chi_R(z_R)$ and still definitely wins with the maximized expected utility from winning. If $\chi_R(z_R) = x_R$, L has an incentive to deviate to choose $z_L = x_L = \chi_L(z_L)$ and has a 50% chance of winning with the maximized expected utility from winning.

- Suppose (B):
  - In cases (1) and (2), L wins with certainty. Here, L has an incentive to deviate to choose $z_L = x_L = \chi_L(z_L)$ and still definitely wins with the maximized expected utility from winning.
  - In cases (3) and (4), they tie. Here, L has an incentive to deviate to choose $z_L = x_L = \chi_L(z_L)$ and has a 50% chance of winning with the maximized expected utility from winning.
  - In case (5), R wins with certainty. Here, R has an incentive to deviate to choose $\chi_R(z'_R)$ such that $\chi_R(z'_R)$ is closer to x_R than $\chi_R(z_R)$ but is still closer to x_m than $\chi_L(z_L)$.

- Suppose (C):
  - In cases (1) and (2), L wins with certainty. Here, L has an incentive to deviate to choose $\chi_L(z'_L)$ such that $\chi_L(z'_L)$ is closer to x_L than $\chi_L(z_L)$ but is still closer to x_m than $\chi_R(z_R)$.

- In cases (3) and (4), they tie. Here, if L deviates to choose $x_m - \left(\chi_L(z_L) - x_m\right)$, the expected utility increases.
- In case (5), R wins with certainty. Here, R has an incentive to deviate to choose $\chi_R(z_R')$ such that $\chi_R(z_R')$ is closer to x_R than $\chi_R(z_R)$ but is still closer to x_m than $\chi_L(z_L)$.

For the same reasons, if $\chi_R(z_R) < x_m$, there is no equilibrium.

2.A.5 Proposition 2.6

Sufficient condition

If the pair of platforms satisfies condition (2.2) and is symmetric, it is in equilibrium. If no one deviates, the payoff for candidate i is

$$(1/2)\left[-u\left(\left|\chi_i(z_i) - x_i\right|\right) - \lambda c\left(\left|z_i - \chi_i(z_i)\right|\right) + b - u\left(\left|\chi_j(z_j) - x_i\right|\right)\right].$$

If candidate i deviates to any policy that diverges from x_m, he/she is certain to lose, and the payoff becomes $-u\left(\left|\chi_j(z_j) - x_i\right|\right)$. The change in payoff from this deviation is

$$(1/2)\left[-u\left(\left|\chi_i(z_i) - x_i\right|\right) - \lambda c\left(\left|z_i - \chi_i(z_i)\right|\right) + b - u\left(\left|\chi_j(z_j) - x_i\right|\right)\right].$$ From (2.2), it is zero, and therefore, there is no profitable deviation that diverges from x_m.

If the candidate deviates to a more moderate platform, say z_i', he/she is certain to win. Suppose that the candidate deviates from z_i to z_i'. After this deviation, the payoff becomes $-u\left(\left|\chi_i(z_i') - x_i\right|\right) - \lambda c\left(\left|z_i' - \chi_i(z_i')\right|\right) + b$. The change in the payoff from this deviation is $-u\left(\left|\chi_i(z_i') - x_i\right|\right) - \lambda c\left(\left|z_i' - \chi_i(z_i')\right|\right) + b - (1/2)\left[-u\left(\left|\chi_i(z_i) - x_i\right|\right) - \lambda c\left(\left|z_i - \chi_i(z_i)\right|\right) + b - u\left(\left|\chi_j(z_j) - x_i\right|\right)\right]$. Since $-u\left(\left|\chi_i(z_i) - x_i\right|\right) - \lambda c\left(\left|z_i - \chi_i(z_i)\right|\right) + b = -u\left(\left|\chi_j(z_j) - x_i\right|\right)$, from (2.2), this can be rewritten as $-u\left(\left|\chi_i(z_i') - x_i\right|\right) - \lambda c\left(\left|z_i' - \chi_i(z_i')\right|\right) + b - \left[-u\left(\left|\chi_i(z_i) - x_i\right|\right) - \lambda c\left(\left|z_i - \chi_i(z_i)\right|\right) + b\right]$. From Corollary 2.3, $-u\left(\left|\chi_i(z_i') - x_i\right|\right) < -u\left(\left|\chi_i(z_i) - x_i\right|\right)$ and $\lambda c\left(\left|z_i' - \chi_i(z_i')\right|\right) > \lambda c\left(\left|z_i - \chi_i(z_i)\right|\right)$. Thus, the change in the payoff from this deviation is negative. Therefore, there is no profitable deviation approaching x_m. As a result, the two platforms satisfy (2.2) and are symmetric. Therefore, this is a state of equilibrium.

Necessary condition

To show the necessary condition, I use a contradiction; that is, if this pair does not satisfy equation (2.2) or is not symmetric, it is not in equilibrium.

First, if the pair of platforms is asymmetric, one candidate loses and the other wins. The winning candidate prefers another platform that has a higher utility, that is, one that approaches his/her own ideal point, x_i, but still wins. Thus, the asymmetric position is not in equilibrium. In what follows, I assume that the candidates' platform positions (and policies they would implement) are symmetric.

Second, if equation (2.2) is not satisfied with a pure strategy, it is not in equilibrium. If $-u\big(\big\|\chi_i(z_i)-x_i\big\|\big)-\lambda c\big(\big\|z_i-\chi_i(z_i)\big\|\big)+b<-u\big(\big\|\chi_j(z_j)-x_i\big\|\big)$ and there is a tie, the candidate has an incentive to deviate to lose. Then, he/she can choose any platform that is worse for the median voter and lose. Before this deviation, the expected utility is $(1/2)\Big[-u\big(\big\|\chi_i(z_i)-x_i\big\|\big)-\lambda c\big(\big\|z_i-\chi_i(z_i)\big\|\big)+b-u\big(\big\|\chi_j(z_j)-x_i\big\|\big)\Big]$. After the deviation, it is $-u\big(\big\|\chi_j(z_j)-x_i\big\|\big)$. Thus, this candidate can increase his/her utility by $(1/2)\Big[u\big(\big\|\chi_i(z_i)-x_i\big\|\big)+\lambda c\big(\big\|z_i-\chi_i(z_i)\big\|\big)-b-u\big(\big\|\chi_j(z_j)-x_i\big\|\big)\Big]$ from this deviation. Since $-u\big(\big\|\chi_i(z_i)-x_i\big\|\big)-\lambda c\big(\big\|z_i-\chi_i(z_i)\big\|\big)+b<-u\big(\big\|\chi_j(z_j)-x_i\big\|\big)$, any candidate will deviate.

If $-u\big(\big\|\chi_i(z_i)-x_i\big\|\big)-\lambda c\big(\big\|z_i-\chi_i(z_i)\big\|\big)+b>-u\big(\big\|\chi_j(z_j)-x_i\big\|\big)$ and there is a tie, then the candidate has an incentive to deviate to be certain of winning. The candidate can move slightly to a platform that is better for the median voter and be certain to win. Assume that the deviation to approach x_m is minor. Before this deviation, the utility is $(1/2)\Big[-u\big(\big\|\chi_i(z_i)-x_i\big\|\big)-\lambda c\big(\big\|z_i-\chi_i(z_i)\big\|\big)+b-u\big(\big\|\chi_j(z_j)-x_i\big\|\big)\Big]$. After the deviation, it is slightly lower than $-u\big(\big\|\chi_i(z_i)-x_i\big\|\big)-\lambda c\big(\big\|z_i-\chi_i(z_i)\big\|\big)+b$. This candidate can increase his/her utility by slightly less than $(1/2)\Big[u\big(\big\|\chi_j(z_j)-x_i\big\|\big)-u\big(\big\|\chi_i(z_i)-x_i\big\|\big)-\lambda c\big(\big\|z_i-\chi_i(z_i)\big\|\big)+b\Big]$ from this deviation. Since $-u\big(\big\|\chi_i(z_i)-x_i\big\|\big)-\lambda c\big(\big\|z_i-\chi_i(z_i)\big\|\big)+b>-u\big(\big\|\chi_j(z_j)-x_i\big\|\big)$ and the policy space is continuous, there exists a platform that can increase the candidate's utility, and hence, either candidate has an incentive to deviate.

Finally, suppose that a candidate chooses a mixed strategy. Denote $\hat{z}_i$ as the platform under which the utilities, should either candidate win, are the same. That is, $-u\big(\big\|\chi_i(\hat{z}_i)-x_i\big\|\big)-\lambda c\big(\big\|\hat{z}_i-\chi_i(\hat{z}_i)\big\|\big)+b=-u\big(\big\|\chi_j(\hat{z}_j)-x_i\big\|\big)$. If this mixed strategy is discrete, a candidate whose mixed strategy includes a more extreme platform than $\hat{z}_i$ has an incentive to deviate slightly to approach the median policy, because the probability of winning increases discretely with only a slight increase in the cost of betrayal and the disutility. If all strategies in a discrete mixed strategy are more moderate than $\hat{z}_i$, a candidate deviates to lose. If a mixed strategy is distributed on a continuous policy space, the probability of winning is zero when a candidate announces the most extreme platform in his/her mixed strategy, given that the two candidates' positions are symmetric. Then, a candidate never chooses such a platform. As a result, equation (2.2) is the necessary condition.

Existence and uniqueness
As I have shown, for both candidates, the policies to be implemented must be symmetric in equilibrium. Here, I show that such a unique, symmetric equilibrium exists. To prove this, I consider the simultaneous and symmetric move of both candidates' policies. From condition (2.2),

$$u\left(\left|\chi_j\left(z_j\right) - x_i\right|\right) - u\left(\left|\chi_i\left(z_i\right) - x_i\right|\right) + b = \lambda c\left(\left|z_i - \chi_i\left(z_i\right)\right|\right) \qquad (2.9)$$

When $z_i = x_i$ for both candidates, $z_i = x_i = \chi_i$. Therefore, the left-hand side of (2.9) is $u(x_R - x_L) + b$. When $\chi_i = x_m$ for both candidates, the left-hand side is b. The value of the left-hand side continuously and strictly decreases to b as $\chi_i(z_i)$ and $\chi_j(z_j)$ approach x_m. When $z_i = x_i = \chi_i$, the cost of betrayal is zero. The right-hand side is positive, continuous, and increasing as $\chi_i(z_i)$ approaches x_m, from Corollary 2.3. There exists a point at which the value of the left-hand side is the same as the cost of betrayal, because I assume $b < \lambda c\left(\left|z_i\left(x_m\right) - x_m\right|\right)$. The left-hand side strictly decreases, and the cost of betrayal increases as $\chi_i(z_i)$ approaches x_m. Hence, this point is unique.

∎

2.A.6 Corollary 2.7

When a candidate chooses $z_i = x_i$, the expected utility is $-u(x_R - x_L)/2 + b/2$, because $u(x_i - x_i) = 0$. When the candidate commits to a policy that is slightly closer to the median policy than his/her ideal policy, and wins, the expected utility is slightly lower than $-\lambda c\left(\left|z_i\left(x_i\right) - x_i\right|\right) + b$. As a result, if $(1/2)\left(u(x_R - x_L) - b\right) < \lambda c\left(\left|z_i\left(x_i\right) - x_i\right|\right)$, the candidate has no incentive to deviate to be certain of winning when he/she chooses $z_i = x_i = \chi_i(z_i)$. Otherwise, candidates have an incentive to commit to implementing a policy that is more moderate than x_i. Therefore, they choose $\{z_L, z_R\}$, which satisfies (2.2).

∎

2.A.7 Proposition 2.8

Fix $\chi_i(z_i)$, and denote it as $\bar{\chi}_i$. Denote $z_i(\bar{\chi}_i)$ as the platform that commits to $\bar{\chi}_i$; that is, $\bar{\chi}_i = \chi_i(z_i(\bar{\chi}_i))$. Differentiate $\lambda c(|z_i(\bar{\chi}_i) - \bar{\chi}_i|)$ by λ, yields

$$c(|z_i(\bar{\chi}_i) - \bar{\chi}_i|) + \lambda c'(|z_i(\bar{\chi}_i) - \bar{\chi}_i|)\frac{\partial z_i(\bar{\chi}_i)}{\partial \lambda} \quad (2.10)$$

Differentiate equation (2.1) by λ, yields

$$1 = \frac{u'(|\bar{\chi}_i - x_i|)c''(|z_i(\bar{\chi}_i) - \bar{\chi}_i|)\frac{\partial z_i(\bar{\chi}_i)}{\partial \lambda}}{c'(|z_i(\bar{\chi}_i) - \bar{\chi}_i|)^2}$$

Thus,

$$\frac{\partial z_i(\bar{\chi}_i)}{\partial \lambda} = \frac{c'(|z_i(\bar{\chi}_i) - \bar{\chi}_i|)^2}{u'(|\bar{\chi}_i - x_i|)c''(|z_i(\bar{\chi}_i) - \bar{\chi}_i|)}$$

Moreover, $\lambda = u'(|\bar{\chi}_i - x_i|)/c'(|z_i(\bar{\chi}_i) - \bar{\chi}_i|)$ in equilibrium, from Lemma 2.1. Substitute these into (2.10). Then, (2.10) becomes

$$c(|z_i(\bar{\chi}_i) - \bar{\chi}_i|) - \frac{c'(|z_i(\bar{\chi}_i) - \bar{\chi}_i|)^2}{c''(|z_i(\bar{\chi}_i) - \bar{\chi}_i|)}$$

which is negative, from Assumption 1.

From condition (2.1), $\lambda c(|z_i - \chi_i(z_i)|) = [c(|z_i - \chi_i(z_i)|) u'(|\chi_i(z_i) - x_i|)/c'(|z_i - \chi_i(z_i)|)]$. If λ goes to infinity, $|z_i - \chi_i(z_i)|$ converges to 0 from Lemma 2.1, and thus, $c(|z_i - \chi_i(z_i)|)/c'(|z_i - \chi_i(z_i)|)$ decreases to zero from Assumption 1. From Lemma 2.5, $|\chi_i(z_i) - x_i|$ does not exceed $|x_m - x_i|$ in equilibrium, and as such, $|\chi_i(z_i) - x_i|$ goes to a certain positive value ($|\chi_i(z_i) - x_i| \in (0, \infty)$) when λ goes to infinity. Therefore, $u'(|\chi_i(z_i) - x_i|)$ goes to a certain positive value when $u''(|\chi_i(z_i) - x_i|) > 0$, which is a constant positive value when $u''(|\chi_i(z_i) - x_i|) = 0$. As a result, the cost of betrayal, $\lambda c(.)$, approaches zero as λ goes to infinity, and then, $b > \lambda c(|z_i(x_m) - x_m|) = 0$ since

$b > 0$. Hence, both candidates choose $\chi_i(z_i) = \chi_j(z_j) = x_m$ from Lemma 2.4.

∎

2.A.8 Proposition 2.9

Sufficient condition and existence

If the pair of platforms satisfies condition (2.4), $\chi_i(z_i) = x_m - (\chi_R(z_R) - x_m) + \epsilon$ if $i = L$, and $x_m + (x_m - \chi_L(z_L)) - \epsilon$ if $i = R$, then this pair is in equilibrium. Candidate j has $\psi_j(z_j, z_i) + b \leq 0$, which means that j does not have an incentive to deviate by approaching the median policy and winning against (or tying with) i. From any other possible deviation, j has the same expected utility, because he/she still loses. Thus, there is no profitable deviation for j. Candidate i has $\psi_i(z_i, z_j) + b \geq 0$, which means that i does not have an incentive to deviate and lose. Moreover, i cannot find a policy that is more extreme than $\chi_i(z_i)$ but that still wins against j, because no such policy exists when $\chi_i(z_i) = x_m - (\chi_R(z_R) - x_m) + \epsilon$ if $i = L$, and $x_m + (x_m - \chi_L(z_L)) - \epsilon$ if $i = R$. Thus, there is no profitable deviation for i. As a result, this is a state of equilibrium. Such an equilibrium exists since $d_i < d_j$ and $d_j > 0$.

Necessary condition

To show the necessary condition, I use a contradiction. First, there is no equilibrium in which the winner commits to implementing a policy that is closer to x_m than the opponent's policy by more than ϵ. That is, $\chi_L(z_L) \geq x_m - (\chi_R(z_R) - x_m) + 2\epsilon$ or $\chi_R(z_R) \leq x_m + (x_m - \chi_L(z_L)) - 2\epsilon$. The winner has an incentive to deviate by choosing $\chi_L(z_L) = x_m - (\chi_R(z_R) - x_m) + \epsilon$ or $\chi_R(z_R) = x_m + (x_m - \chi_L(z_L)) - \epsilon$, that is, committing to a policy that is closer to his/her ideal policy and still wins. In what follows, I exclude such cases.

Second, because $d_i < d_j$, there is no possibility of satisfying both $\psi_i(z_i, z_j) + b = 0$ and $\psi_j(z_j, z_i) + b = 0$ at the same time with $|\chi_i(z_i) - x_m| = |\chi_j(z_j) - x_m|$. Therefore, no symmetric equilibrium exists in which both candidates have the same probability of winning, according to Proposition 2.6.

Third, suppose that $\chi_i(z_i) = x_m - (\chi_R(z_R) - x_m) + \epsilon$ if $i = L$, and $x_m + (x_m - \chi_L(z_L)) - \epsilon$ if $i = R$. Because $d_i < d_j$, and ϵ is very small, it is not possible to satisfy both $\psi_i(z_i, z_j) + b \leq 0$ and $\psi_j(z_j, z_i) + b > 0$ at the same time, or to satisfy both $\psi_i(z_i, z_j) + b < 0$ and $\psi_j(z_j, z_i) + b \geq 0$ at the same time. If $\psi_i(z_i, z_j) + b < 0$ and $\psi_j(z_j, z_i) + b < 0$, this is not a state of equilibrium because i wants to deviate to lose. Furthermore, if $\psi_i(z_i, z_j) + b > 0$ and $\psi_j(z_j, z_i) + b > 0$, this is not in equilibrium because j wants to win with certainty (or with a 50% chance) by approaching the median policy.

As a result, condition (2.4), along with $\chi_i(z_i) = x_m - (\chi_R(z_R) - x_m) + \epsilon$ if $i = L$, and $x_m + (x_m - \chi_L(z_L)) - \epsilon$ if $i = R$ is the necessary condition.

■

2.A.9 Corollary 2.10

Suppose that the two candidates are originally symmetric (i.e., they have symmetric cost and disutility functions, and their ideal policies are equidistant from the median policy), and they announce symmetric platforms. Thus, they will implement χ_L and χ_R, which are also symmetric. Moreover, both candidates initially have $\Psi_i(z_i, z_j) + b = 0$. Then, consider that R becomes more extreme than L. (i.e., x_R increases). If $\psi_R(z_R, z_L)$ decreases, then from Proposition 2.9, a more moderate L wins against a more extreme R, with certainty.

Denote $z_R(\chi_R) = \chi_R^{-1}(\chi_R)$, which is the platform committing a candidate to χ_R. Fix χ_L and χ_R, and assume that χ_L and χ_R are symmetric. Differentiate $u(x_R - \chi_L) - u(x_R - \chi_R) - \lambda c(\chi_R - z_R(\chi_R))$ by x_R, which gives $u'(x_R - \chi_L) - u'(x_R - \chi_R) + \lambda c'(\chi_R - z_R(\chi_R))(\partial z_R(\chi_R) / \partial x_R)$. Now, differentiate equation (2.1) by x_R. Then,

$$\frac{\partial z_R(\chi_R)}{\partial x_R} = -\frac{u''(x_R - \chi_R)c'(\chi_R - z_R(\chi_R))}{u'(x_R - \chi_R)c''(\chi_R - z_R(\chi_R))} < 0$$

Moreover, $\lambda = u'(x_R - \chi_R) / c'(\chi_R - z_R(\chi_R))$ in equilibrium, from Lemma 2.1. After substituting these into the above equation, we have

$$u'(x_R - \chi_L) - u'(x_R - \chi_R) - \frac{u''(x_R - \chi_R)c'(\chi_R - z_R(\chi_R))}{c''(\chi_R - z_R(\chi_R))} \quad (2.11)$$

If (2.11) is negative, $\psi_R(z_R, z_L)$ becomes lower than $\psi_L(z_L, z_R)$ when R becomes more extreme, which means that R will lose. Equation (2.11) is negative if

$$\frac{u'(x_R - \chi_L) - u'(x_R - \chi_R)}{u''(x_R - \chi_R)} < \frac{c'(\chi_R - z_R(\chi_R))}{c''(\chi_R - z_R(\chi_R))} \quad (2.12)$$

Since this was originally $\psi_i(z_i, z_j) + b = 0$ and $\lambda = u'(x_R - \chi_R)$ $/ c'(\chi_R - z_R(\chi_R))$,

$$\frac{u(x_R - \chi_L) - u(x_R - \chi_R)}{u'(x_R - \chi_R)} < \frac{c(\chi_R - z_R(\chi_R))}{c'(\chi_R - z_R(\chi_R))} \qquad (2.13)$$

From Assumption 1, $c'(\chi_R - z_R(\chi_R)) / c''(\chi_R - z_R(\chi_R)) > c(\chi_R - z_R(\chi_R)) / c'(\chi_R - z_R(\chi_R))$; that is, the right-hand side of (2.12) is higher than the right-hand side of (2.13). If

$$\frac{u'(x_R - \chi_L) - u'(x_R - \chi_R)}{u''(x_R - \chi_R)} \leq \frac{u(x_R - \chi_L) - u(x_R - \chi_R)}{u'(x_R - \chi_R)}$$

(2.12) holds. This equation can be changed to

$$\frac{u'(x_R - \chi_L)}{u''(x_R - \chi_R)} - \frac{u(x_R - \chi_L)}{u'(x_R - \chi_R)} \leq \frac{u'(x_R - \chi_R)}{u''(x_R - \chi_R)} - \frac{u(x_R - \chi_R)}{u'(x_R - \chi_R)}$$

If $x_R - \chi_L = x_R - \chi_R$, both sides are the same. If $x_R - \chi_L$ increases, the left-hand side decreases or does not change. The reason is as follows. Differentiate the left-hand side with respect to $x_R - \chi_L$, which gives $u''(x_R - \chi_L) / u''(x_R - \chi_R) - u'(x_R - \chi_L) / u'(x_R - \chi_R)$. This value is non-positive because $u'(x_R - \chi_R) / u''(x_R - \chi_R) \leq u'(x_R - \chi_L) / u''(x_R - \chi_L)$ when $x_R - \chi_L > x_R - \chi_R$ from Assumption 2. As a result, the left-hand side of (2.12) is lower than or the same as the left-hand side of (2.13). Therefore, (2.12) holds, and the candidate with a lower $|x_i - x_m|$ (L in this case) wins.

Consider $|x_i - x_m| < |x_j - x_m|$. Then, candidate i wins with certainty and, for candidate i, $\psi_i(z_i, z_j) + b$ is not negative. That is, $-u(|\chi_i(z_i) - x_i|) - \lambda_i c(|z_i - \chi_i(z_i)|) + b \geq -u(|\bar{\chi}_j - x_i|)$, where $\bar{\chi}_j$ satisfies $|x_m - \chi_i(z_i)| = |x_m - \bar{\chi}_j|$. Since $|x_i - x_m| < |x_j - x_m|$, $-u(|\bar{\chi}_j - x_i|) > -u(|\chi_i(z_i) - x_j|)$, and then, $-u(|\chi_i(z_i) - x_i|) - \lambda_i c(|z_i - \chi_i(z_i)|) + b > -u(|\chi_i(z_i) - x_j|)$. The left-hand side is the (expected) utility for candidate i (utility from winning) and the right-hand side is the (expected) utility for candidate j (utility from losing).

When the candidates have a linear utility function, $u'(x_R - \chi_L) = u'(x_R - \chi_R)$ and $\partial z_R(\chi_R) / \partial x_R = 0$, the change in both sides of the first-order condition is zero as x_R changes. Thus, regardless of the position of the candidates, they still tie if they have an incentive to approach x_m more than x_i. If both candidates (or either) do not have such an incentive and choose $z_i = x_i = \chi_i(z_i)$, a more moderate candidate is certain to win.

∎

54   *Electoral promises as a commitment device*

2.A.10 Corollary 2.11

Suppose that the two candidates are originally symmetric (i.e., they have symmetric cost and disutility functions, and their ideal policies are equidistant from the median policy), and they announce symmetric platforms. Thus, they will implement χ_L and χ_R, which are also symmetric. Moreover, originally, both candidates have $\psi_i(z_i, z_j) + b = 0$. Then, consider that R's relative importance of betrayal, λ, becomes lower than that of L's (i.e., λ_R decreases). If $\psi_R(z_R, z_L)$ decreases, from Proposition 2.9, L, which has a higher λ_L, always wins against R, which has a lower λ_R.

Now, fix χ_L and χ_R, and assume that χ_L and χ_R are symmetric. Differentiate $u(x_R - \chi_L) - u(x_R - \chi_R) - \lambda_R c(\chi_R - z_R(\chi_R))$ with respect to λ_R, which yields $-c(\chi_R - z_R(\chi_R)) + \lambda_R c'(\chi_R - z_R(\chi_R))(\partial z_R(\chi_R) / \partial \lambda_R)$. Differentiate equation (2.1) with respect to λ_R. Then,

$$\frac{\partial z_R(\chi_R)}{\partial \lambda_R} = \frac{c'(\chi_R - z_R(\chi_R))^2}{u'(x_R - \chi_R)c''(\chi_R - z_R(\chi_R))} > 0$$

Moreover, $\lambda_R = u'(x_R - \chi_R) / c'(\chi_R - z_R(\chi_R))$ in equilibrium, from Lemma 2.1. Substituting these values in the above equation, it becomes $-c(\chi_R - z_R(\chi_R)) + c'(\chi_R - z_R(\chi_R))^2 / c''(\chi_R - z_R(\chi_R))$, which is positive, from Assumption 1. As a result, the candidate with a higher λ_i (L, in this case) always wins.

For candidate i, the utility when he/she wins is higher than or equal to the utility when the opponent wins; that is, $-u(|\chi_i(z_i) - x_i|) - \lambda_i c(|z_i - \chi_i(z_i)|) + b \geq -u(|\bar{\chi}_j - x_i|)$, where $\bar{\chi}_j$ satisfies $|x_m - \chi_i(z_i)| = |x_m - \bar{\chi}_j|$. Since $|x_i - x_m| = |x_j - x_m|$, $-u(|\bar{\chi}_j - x_i|) = -u(|\chi_i(z_i) - x_j|)$. Therefore, $-u(|\chi_i(z_i) - x_i|) - \lambda_i c(|z_i - \chi_i(z_i)|) + b \geq -u(|\chi_i(z_i) - x_j|)$. The left-hand side is the (expected) utility of i (from winning), and the right-hand side is the (expected) utility of j (from losing).

∎

2.A.11 Corollary 2.12

Suppose that two candidates are originally symmetric (i.e., they have symmetric cost and disutility functions, and their ideal policies are equidistant from the median policy), and they announce symmetric platforms. Thus, they will implement χ_L and χ_R, which are also symmetric. Moreover, both candidates initially have $\psi_i(z_i, z_j) + b = 0$. Then, consider that R becomes more policy motivated than L (i.e., β_R

increases). If $\psi_R(z_R, z_L)$ decreases, then from Proposition 2.9, the less policy-motivated L always wins against a more policy-motivated R.

Now, fix χ_L and χ_R and assume that χ_L and χ_R are symmetric. Differentiate $\beta_R u(x_R - \chi_L) - \beta_R u(x_R - \chi_R) - \lambda c(\chi_R - z_R(\chi_R))$ with respect to β_R, which yields $u(x_R - \chi_L) - u(x_R - \chi_R) - \lambda c'(\chi_R - z_R(\chi_R)) \cdot (\partial z_R(\chi_R)/\partial \beta_R)$ Differentiate equation (2.1), $\lambda = \beta_R u'(x_R - \chi_R)/ c'(\chi_R - z_R(\chi_R))$, with respect to β_R. We get

$$\frac{\partial z_R(\chi_R)}{\partial \beta_R} = -\frac{c'(\chi_R - z_R(\chi_R))}{\beta_R c''(\chi_R - z_R(\chi_R))} < 0.$$

Moreover, $\lambda = \beta_R u'(x_R - \chi_R)/c'(\chi_R - z_R(\chi_R))$ in equilibrium, from Lemma 2.1. On substituting these into the above equation, we have

$$u(x_R - \chi_L) - u(x_R - \chi_R) - u'(x_R - \chi_R)\frac{c'(\chi_R - z_R(\chi_R))}{c''(\chi_R - z_R(\chi_R))} \qquad (2.14)$$

Since this was originally $\psi_i(z_i, z_j) + b = 0$ and $\lambda = \beta_R u'(x_R - \chi_R) / c'(\chi_R - z_R(\chi_R))$, (2.13) is satisfied in the proof of Corollary 2.10. From Assumption 1, $c'(\chi_R - z_R(\chi_R))/ c''(\chi_R - z_R(\chi_R)) > c(\chi_R - z_R(\chi_R))/ c'(\chi_R - z_R(\chi_R))$, and thus (2.14) is negative. Therefore, the candidate with a lower β_i (L, in this case) always wins.

Consider $\beta_i < \beta_j$. Then, i always wins, which means that in equilibrium, $\psi_i(z_i, z_j) + b = 0$ is not negative for i. That is, $-\beta_i u(|\chi_i(z_i) - x_i|) - \lambda c(|z_i - \chi_i(z_i)|) + b \geq -\beta_i u(|\bar{\chi}_j - x_i|)$, where $\bar{\chi}_j$ satisfies $|x_m - \chi_i(z_i)| = |x_m - \bar{\chi}_j|$. Note that if $\psi_i(z_i, z_j) + b < 0$, i does not have an incentive to win in equilibrium. Since $\beta_i < \beta_j$, $-\beta_i u(|\bar{\chi}_j - x_i|) > -\beta_j u(|\chi_i(z_i) - x_j|)$, and hence, $-\beta_i u(|\chi_i(z_i) - x_i|) - \lambda c(|z_i - \chi_i(z_i)|) + b \geq -\beta_j u(|\chi_i(z_i) - x_j|)$. The left-hand side is the (expected) utility of i (from winning), and the right-hand side is the (expected) utility of j (from losing).

■

Notes

1 This chapter is revised version of Asako (2015a).
2 It is well known that there are many equilibria in a multi-candidate competition, assuming completely binding platforms (see Adams et al., 2005).

56 *Electoral promises as a commitment device*

Such problems arise even when analyzing partially binding platforms. Moreover, in past studies based on a citizen – candidate framework, equilibria tended to feature one or two candidates only. Thus, we consider only two potential candidates to simplify the analysis.

3 If $u(.)$ is linear and λ is sufficiently low, a candidate promises the ideal policy as a platform from Corollary 2.7.

4 See "Abe Stumbles on Japan," *The Economist*, July 30, 2007.

5 In particular, assume that there exist discrete policies, χ_i and χ_j, such that $\left(\psi_i \left(z_i, z_j \right) + b = 0 \right)$ and $\left| \chi_i - x_m \right| = \left| \chi_j - x_m \right|$, for $i = L$ and R, and $i \neq L$. Therefore, d_i exists, as defined by (2.3).

6 Robinson and Verdier (2013) and Keefer and Vlaicu (2008) study clientelism. Gehlbach et al. (2010) analyze transition economies, especially Russia, in which platforms are non-binding, whereas platforms are completely binding in mature democracies.

7 Cox and McCubbins (1994), Aldrich (1997), Djankov et al. (2003), and Reinikka and Svensson (2005) indicate these points.

8 Zupan (1990), Carey (1994), and Figlio (2000) demonstrate the last-term problem, which describes how a retirement decision induces political shirking.

9 Condition (2.7) ((2.8)) means that, for j (i), the expected utility when both candidates run is greater than or equal to the utility when only opponent i (j) runs. Suppose that (2.7) is satisfied with inequality, but (2.8) is not satisfied. Then, first, if only i runs, j has an incentive to deviate by running. Second, if only j runs, i does not have an incentive to run (and j does not have an incentive to deviate). Third, if both candidates run, i has an incentive to deviate by not running. Thus, there exists only one equilibrium, namely, when only j runs. The inverse case is also true.

10 To be precise, (i) if both (2.7) and (2.8) are satisfied with equality, or (ii) if one of them is satisfied with equality and the other is not, there exists an equilibrium in which either of the two potential candidates runs.

11 The SDPJ formed a coalition government with the LDP from 1994 to 1996; however, it branched out as an individual, smaller party after 1996.

Bibliography

Adams, J., S. Merrill III, and B. Grofman, 2005, *A Unified Theory of Party Competition*, Cambridge: Cambridge University Press.

Aldrich, J., 1997, *Why Parties? The Origin and Transformation of Political Parties in America*, Chicago: University of Chicago Press.

Asako, Y., 2015a, "Partially Binding Platforms: Campaign Promises vis-a-vis Cost of Betrayal," *Japanese Economic Review* 66(3), pp.322–353. (https://doi.org/10.1111/jere.12053)

Calvert, R., 1985, "Robustness of the Multidimensional Voting Model: Candidate Motivations, Uncertainty, and Convergence," *American Journal of Political Science* 29(1), pp. 69–95. (https://doi.org/10.2307/2111217)

Carey, J., 1994, "Political Shirking and the Last Term Problem: Evidence from a Party-Administered Pension System," *Public Choice* 81(1–2), pp. 1–22. (https://doi.org/10.1007/BF01053263)

Cox, G., and M. McCubbins, 1994, *Legislative Leviathan: Party Government in the House*, Los Angeles: University of California Press.

Djankov, S., C. McLiesh, T. Nenova, and A. Shleifer, 2003, "Who Owns the Media?," *Journal of Law and Economics* 46(2), pp. 341–382. (https://doi.org/10.1086/377116)

Downs, A., 1957, *An Economic Theory of Democracy*, New York: Harper and Row.

Ferraz, C., and F. Finan, 2011, "Electoral Accountability and Corruption: Evidence from the Audits of Local Governments," *American Economic Review*, 101(4), pp. 1274–1311. (https://doi.org/10.1257/aer.101.4.1274)

Figlio, D., 2000, "Political Shirking, Opponent Quality, and Electoral Support," *Public Choice* 103, pp. 271–284. (https://doi.org/10.1023/A:1005051613015)

Gehlbach, S., K. Sonin, and E. Zhuravskaya, 2010, "Businessman Candidate," *American Journal of Political Science* 54(3), pp. 718–736. (https://doi.org/10.1111/j.1540-5907.2010.00456.x)

Ihori, T., and T. Doi, 1998, *Nihon-seizi no Keizai Bunseki* (*Economic Analysis on Japanese Politics*), Tokyo: Bokutakusya.

Kanai, T., 2003, *Manifesuto: Atarashii Seiji no Chouryuu* [*Manifesto: New Political Trend*], Tokyo: Kobunsha.

Keefer, P., 2007, "Clientelism, Credibility, and the Policy Choices of Young Democracies," *American Journal of Political Science* 51(4), pp. 804–821. (https://doi.org/10.1111/j.1540-5907.2007.00282.x)

Keefer, P., and R. Vlaicu, 2008, "Democracy, Credibility, and Clientelism," *Journal of Law, Economics, & Organization* 24(2), pp. 371–406. (https://doi.org/10.1093/jleo/ewm054)

Mulgan, A., 2002, *Japan's Failed Revolution: Koizumi and the Politics of Economic Reform*, Canberra: Asia Pacific Press.

Osborne, M., and A. Slivinski, 1996, "A Model of Political Competition with Citizen-Candidates," *The Quarterly Journal of Economics* 111(1), pp. 65–96. (https://doi.org/10.2307/2946658)

Reinikka, R., and J. Svensson, 2005, "Fighting Corruption to Improve Schooling: Evidence from a Newspaper Campaign in Uganda," *Journal of the European Economic Association* 3(2–3), pp. 259–267. (https://doi.org/10.1162/jeea.2005.3.2-3.259)

Robinson, J., and T. Verdier, 2013, "The Political Economy of Clientelism," *Scandinavian Journal of Economics* 115(2), pp. 260–291. (https://doi.org/10.1111/j.1467-9442.2013.12010.x)

Sartori, G., 1976, *Parties and Party System*, Cambridge: Cambridge University Press.

Zupan, M., 1990, "The Last Period Problem in Politics: Do Congressional Representatives Not Subject to a Reelection Constraint Alter Their Voting Behavior?," *Public Choice* 65(2), pp. 167–180. (https://doi.org/10.1007/BF00123797)

3   Electoral promises as a signal

3.1 Introduction

This chapter analyzes the signaling role of campaign promises by extending the model with partially binding platforms discussed in Chapter 2.[1] In particular, I introduce asymmetric information by assuming that candidate policy preferences are private information. One candidate's ideal policy is to the left of the median policy, while that of the other candidate is to the right. Each candidate is one of two types – moderate or extreme – and the moderate type's ideal policy is closer to the median policy than that of the extreme type. A candidate knows his/her own type, but voters and the opponent do not. In the remainder of this chapter, I refer to an extreme type as "he" and a moderate type as "she." Note that both candidates have the same cost of betrayal and the degree of policy motivation ($\lambda = \beta = 1$), so only policy preferences are possibly asymmetric.

Most studies on the two-candidate model of political competition show a politician's convergence with the ideal policy of the median voter. However, in real-world elections, politicians are frequently polarized. This chapter provides one possible reason why politicians are polarized and shows that an extreme candidate has a higher probability of winning than does a moderate candidate, *although the extreme candidate will implement a more extreme policy.* The important reason for the extreme candidate's higher probability of winning is that he has a stronger incentive to prevent an opponent from winning because his ideal policy is further from the opponent's policy than is that of a moderate candidate. The model in this chapter describes this incentive for an extreme candidate by introducing partially binding platforms and uncertainty about a candidate's preference.

3.2 The model

3.2.1 Setting

The policy space is $\mathbb{R}$.[2] There is a continuum of voters, and their ideal policies are distributed on some interval of $\mathbb{R}$. The distribution function is continuous and strictly increasing, so there is a unique median voter's ideal policy, x_m. There are two candidates, L and R, and each candidate is one of two types: moderate or extreme. Let x_i^M and x_i^E denote the respective ideal policies for the *moderate* and *extreme types*, where $i = L$ or R, and $x_L^E < x_L^M < x_m < x_R^M < x_R^E$. The superscripts M and E represent a moderate or extreme type, respectively, and the moderate type's ideal policy is closer to the median policy. Assume $x_m - x_L^t = x_R^t - x_m$ for $t = M$ or E. That is, the ideal policies of the same type are equidistant from the median policy. A candidate knows his/her own type, but voters and the opponent are uncertain about the candidate's type. For both candidates, $p^M \in (0,1)$ is the prior probability that the candidate is a moderate type. Thus, the prior probability that the candidate is an extreme type is $p^E = 1 - p^M$.

After the types of candidates are decided, each candidate announces a platform, denoted by $z_i^t \in \mathbb{R}$, where $i = L$ or R and $t = M$ or E. On the basis of these platforms, voters decide on a winner according to a majority voting rule. After an election, the winning candidate chooses an implemented policy, denoted by χ_i^t, where $i = L$ or R and $t = M$ or E. As in Chapter 2, the policy to be implemented will lie somewhere between the platform policy and the ideal policy, as shown in Figure 1.1 in Chapter 1.

If the implemented policy is different from the candidate's ideal policy, all candidates — both winner and loser — experience disutility. This disutility is represented by $-u\big(\big|\chi - x_i^t\big|\big)$, where $i = L$ or R, $t = M$ or E, and χ is the policy implemented by the winner. Assume that $u(.)$ satisfies $u(0) = 0$, $u'(0) = 0$, $u'(d) > 0$, and $u''(d) > 0$ when $d > 0$. Unlike in Chapter 2, I do not consider linear utility here. If the implemented policy is not the same as the platform, the winning candidate needs to pay a cost of betrayal. The function describing the cost of betrayal is $c\big(\big|z_i^t - \chi\big|\big)$. Assume that $c(.)$ satisfies $c(0) = 0$, $c'(0) = 0$, $c'(d) > 0$, and $c''(d) > 0$ when $d > 0$. The loser does not pay a cost. After an election, the winning candidate chooses a policy that maximizes $-u\big(\big|\chi - x_i^t\big|\big) - c\big(\big|z_i^t - \chi\big|\big)$. That is, the winner chooses

$$\chi_i^t(z_i) = \operatorname{argmax}_\chi -u\big(\big|\chi - x_i^t\big|\big) - c\big(\big|z_i^t - \chi\big|\big).$$

I also assume that $c'(d)/c(d)$ strictly decreases as d increases. That is, Assumption 1 introduced in Chapter 2 holds. I also suppose the following assumption, which is similar to Assumption 2 in Chapter 2.

Assumption 3

$u'(d)/u(d)$, and $u''(d)/u'(d)$ strictly decrease as d increases.

This assumption means that the relative marginal disutility decreases as $\left|\chi - x_i^t\right|$ increases, and the Arrow-Pratt measure of absolute risk aversion is decreasing in $\left|\chi - x_i^t\right|$. For example, if the function is a monomial, this assumption holds, and many polynomial functions satisfy the assumptions as well.

Upon observing a platform, the utility of voter n when candidate i of type t wins is $-u\left(\left|\chi_i^t(z_i) - x_n\right|\right)$. Assume that $u(.)$ satisfies $u'(d) > 0$ when $d > 0$. Let $p_i(t \mid z)$ denote the voters' revised beliefs that candidate i is of type t upon observing platform z. The expected utility of voter n when the winner is candidate i, who promises z_i, is $-p_i(M \mid z_i)u\left(\left|\chi_i^M(z_i) - x_n\right|\right) - (1 - p_i(M \mid z_i))u\left(\left|\chi_i^E(z_i) - x_n\right|\right)$. Voters vote sincerely, which means that they vote for the most preferred candidate, and weakly dominated strategies are ruled out. Assume that all voters and the opponent have the same beliefs about a candidate's type.

Let $\pi_i^t\left(z_i^t, z_j^s\right)$ denote the probability of candidate i of type t winning against opponent j of type s, given z_i^t and z_j^s. Let $F_i^t(.)$ denote the distribution function of the mixed strategy chosen by candidate i of type t. The expected utility of candidate i of type t who promises z_i^t is

$$\sum_{s=M,E}\left[p^s\int_{z_j^s}\pi_i^t\left(z_i^t, z_j^s\right)dF_j^s\left(z_j^s\right)\right]\left[-u\left(\left|\chi_i^t\left(z_i^t\right) - x_i^t\right|\right) - c\left(\left|z_i^t - \chi_i^t\left(z_i^t\right)\right|\right)\right]$$

$$-\sum_{s=M,E}p^s\int_{z_j^s}\left(1 - \pi_i^t\left(z_i^t, z_j^s\right)\right)u\left(\left|\chi_j^s\left(z_j^s\right) - x_i^t\right|\right)dF_j^s\left(z_j^s\right) \qquad (3.1)$$

where $i, j = L, R$ and $t = M, E$. The first term indicates when the candidate defeats each type of opponent. The second term indicates when the candidate loses to each type of opponent. To simplify, I do not introduce a benefit from holding office (i.e., $b = 0$) in a candidate's utility. A benefit from holding office would not change my results significantly when the benefit is small. Here, candidates approach the median policy more closely, but the main characteristics of the equilibria do not change. Therefore, I simply assume that the benefit from holding office is zero. However, if the benefits from holding office are great, candidates' implemented policies converge to the median policy regardless of type like as Lemma 2.4 in Chapter 2. This is what Huang (2010) shows by introducing sufficiently high benefits from holding office, which compensate for all disutility resulting from the policy and the cost of betrayal.

In summary, the timing of events is as follows. Note that this chapter does not consider candidates' decisions to run.

1   Nature decides each candidate's type, and a candidate knows his/her own type.
2   The candidates announce their platforms.
3   Voters vote.
4   The winning candidate chooses which policy to implement.

In what follows, I concentrate on a symmetric, pure-strategy, perfect Bayesian equilibrium consisting of strategies and beliefs.

3.2.2 *Policy implemented by the winner*

Following an election, the winning candidate implements a policy that maximizes his/her utility following a win, $-u\left(\left\|\chi_i^t\left(z_i^t\right)-x_i^t\right\|\right)-c\left(\left\|z_i^t-\chi_i^t\left(z_i^t\right)\right\|\right)$ like as Lemma 2.1 in Chapter 2.

Lemma 3.1

The implemented policy $\chi_i^t(z)$ satisfies $u'\left(\left\|\chi_i^t(z)-x_i^t\right\|\right)=c'\left(\left\|z-\chi_i^t(z)\right\|\right)$, for $\chi_i^t(z)\in\left(x_i^t,z\right)$ when $z>x_i^t$, and $\chi_i^t(z)\in\left(z,x_i^t\right)$ when $z<x_i^t$.

When the platform differs from the ideal policy, the implemented policy must differ from the platform or the ideal policy since it is decided by the winner (who no longer cares about the loser's platform, but cares about the cost of betrayal) after an election. If voters know the candidate's type (ideal policy), they can also know the future implemented policy by observing the platform. However, with asymmetric information, they may not know the candidate's type. Here, the median voter x_m is pivotal. Thus, if the candidate is more attractive to the median voter than is the opponent, this candidate is certain to win.

3.3 Pooling equilibrium

3.3.1 *Definition and proposition*

If both types of opponent announce the same platform z_j (i.e., a pooling strategy), the expected utility of candidate i of type t when the opponent wins is $-p^M u\left(\left|\chi_j^M(z_j)-x_i^t\right|\right)-\left(1-p^M\right)u\left(\left|\chi_j^E(z_j)-x_i^t\right|\right)$, where $i,j=L,R$, $i\neq j$, and $t=M,E$. The utility of candidate i of type t after he/she wins is $-u\left(\left|\chi_i^t\left(z_i^t\right)-x_i^t\right|\right)-c\left(\left|z_i^t-\chi_i^t\left(z_i^t\right)\right|\right)$. If the utility when candidate i of type t wins is strictly lower than the expected utility when his/her opponent wins $\left(-p^M u\left(\left|\chi_j^M(z_j)-x_i^t\right|\right)-\left(1-p^M\right)u\left(\left|\chi_j^E(z_j)-x_i^t\right|\right)>-u\left(\left|\chi_i^t\left(z_i^t\right)-x_i^t\right|\right)-c\left(\left|z_i^t-\chi_i^t\left(z_i^t\right)\right|\right)\right)$, candidate i prefers the opponent winning to him/herself winning. On the other hand, in the inverse case (i.e., $-p^M u\left(\left|\chi_j^M(z_j)-x_i^t\right|\right)-\left(1-p^M\right)u\left(\left|\chi_j^E(z_j)-x_i^t\right|\right)<-u\left(\left|\chi_i^t\left(z_i^t\right)-x_i^t\right|\right)-c\left(\left|z_i^t-\chi_i^t\left(z_i^t\right)\right|\right)$, candidate i prefers to win.

In a pooling equilibrium, a moderate type chooses a platform z_i^{M*} such that she is indifferent between winning and losing. That is, the above two expected utilities are the same for a moderate type:

$$
\begin{aligned}
&-p^M u\left(\left|\chi_j^M\left(z_j^{M*}\right)-x_i^M\right|\right)-\left(1-p^M\right)u\left(\left|\chi_j^E\left(z_j^{M*}\right)-x_i^M\right|\right)\\
&=-u\left(\left|\chi_i^M\left(z_i^{M*}\right)-x_i^M\right|\right)-c\left(\left|z_i^{M*}-\chi_i^t\left(z_i^{M*}\right)\right|\right)
\end{aligned}
\tag{3.2}
$$

where z_L^{M*} and z_R^{M*} are symmetric $\left(\left|x_m-z_L^{M*}\right|=\left|x_m-z_R^{M*}\right|\right)$. An extreme type mimics a moderate type by choosing the same platform.[3]

Definition 3.2

In a pooling equilibrium, z_i^{M*} is chosen by a candidate regardless of his/her type.

Then, a pooling equilibrium exists if the prior belief that a candidate is of a moderate type, p^M, is sufficiently high. The parameter, $\bar{p}$, will be defined later (Equation (3.6)).

Proposition 3.3

Suppose that the off-path beliefs of voters are $p_i(M|z_i)=0$. If $p^M \geq \bar{p}$, the pooling equilibrium defined in Definition 3.2 exists.

Proof: See Appendix 3.A.1.

3.3.2   *An extreme type's choice*

Intuitively, a moderate type is indifferent between winning and losing, with z_i^{M*} so she does not deviate. On the other hand, this subsection shows that an extreme type prefers his winning to the opponent winning, with z_i^{M*}.

Let $z_i^t(z_j)$ denote the *cut-off* platform, where the utility when candidate i wins and the expected utility when opponent j wins are the same for type t candidate i, given the opponent's pooling strategy, z_j. That is, a candidate is indifferent between winning and losing:

$$
\begin{aligned}
&-p^M u\left(\left\|\chi_j^M(z_j)-x_i^t\right\|\right)-\left(1-p^M\right)u\left(\left\|\chi_j^E(z_j)-x_i^t\right\|\right)\\
&=-u\left(\left\|\chi_i^t\left(z_i^t(z_j)\right)-x_i^t\right\|\right)-c\left(\left\|z_i^t(z_j)-\chi_i^t\left(z_i^t(z_j)\right)\right\|\right)
\end{aligned}
\tag{3.3}
$$

For example, if the opponent chooses z_j^{M*}, z_i^{M*} is the cut-off platform of moderate candidate i $\left(z_i^M\left(z_j^{M*}\right)=z_i^{M*}\right)$. If a candidate approaches the median policy, the disutility after winning and the cost of betrayal increase. Thus, if type t candidate i announces a platform that is further from his/her ideal policy (i.e., more moderate) than $z_i^t(z_j)$, then his/her utility after winning is lower than the expected utility when the opponent wins, and vice versa.

In this chapter, when I use the term "more moderate platform," this means that "this platform is further from the candidate's ideal policy." In Figure 3.1, $z_R^E(z_L)$ is further from x_R^E and x_R^M than $z_R^M(z_L)$. Therefore, $z_R^E(z_L)$ is "more moderate" than $z_R^M(z_L)$. Note also that "approaching the median policy" means that a candidate announces a platform such that an implemented policy given this platform approaches the median policy.

Note that, first, a more moderate platform does not mean a more moderate implemented policy, given this platform. As shown in Figure 3.1, since an extreme type will betray his platform to a greater extent than will a moderate type, the extreme type's implemented policy, $\chi_R^E\left(z_R^E(z_L)\right)$, is more extreme than that of the moderate type, $\chi_R^M\left(z_R^M(z_L)\right)$. Note also that a more moderate platform may not mean that this platform is closer to the median policy, because there is a possibility that a platform encroaches on the opponent's side of

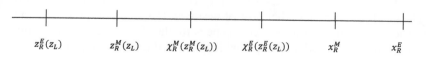

Figure 3.1 Lemma 3.4.
While an extreme type's implemented policy given the cut-off platform $\chi_R^E\left(z_R^E(z_L)\right)$ is more extreme than a moderate type's one $\chi_R^M\left(z_R^M(z_L)\right)$, an extreme type's cut-off platform $z_R^E(z_L)$ is more moderate than a moderate type's one $z_R^E(z_L)$.

the policy space (i.e., $z_R^t < x_m < z_L^t$) like as the model in Chapter 2. The justifications discussed in Subsection 2.3.3 can be also applied here.

If an extreme type's $z_i^E(z_j)$ is always more moderate than a moderate type's $z_i^M(z_j)$, given any z_j $\left(z_L^M(z_R) < z_L^E(z_R) \text{ and } z_R^E(z_L) < z_R^M(z_L) \right)$, the extreme type prefers his winning to the opponent winning when both candidates announce z_i^{M*} and z_j^{M*}. The following lemma shows that this is always true. Figure 3.1 demonstrates the lemma in the case of R.

Lemma 3.4

Suppose that an opponent announces the same platform z_j regardless of type. Given any p^M, (i) an extreme type's cut-off platform is more moderate than that of a moderate type $\left(z_L^M(z_R) < z_L^E(z_R) \text{ and } z_R^E(z_L) < z_R^M(z_L) \right)$, but (ii) an extreme type's implemented policy, given the cut-off platform, is more extreme than that of a moderate type $\left(\chi_L^E \left(z_L^E(z_R) \right) < \chi_L^M \left(z_L^M(z_R) \right) \text{ and } \chi_R^E \left(z_R^E(z_L) \right) > \chi_R^M \left(z_R^M(z_L) \right) \right)$.
Proof: See Appendix 3.A.2.

Lemma 3.4 shows the following two facts.

1   An extreme type has an incentive to announce a more moderate platform than does a moderate type.
2   A moderate type has an incentive to commit to implementing a more moderate implemented policy than does an extreme type.

If a candidate approaches the median policy and wins against the opponent with certainty, this candidate will pay a cost of betrayal with certainty. This marginal cost of approaching the median policy depends on the cost of betrayal, $-c\left(\left| z_i^t - \chi_i^t \left(z_i^t \right) \right| \right)$. On the other hand, this candidate can avoid the opponent's victory and decrease his/her expected disutility from the implemented policy. This marginal benefit depends on the difference in the (expected) disutilities when the candidate wins and when the opponent wins,

$$p^M u\left(\left| \chi_j^M \left(z_j^M \right) - x_i^t \right| \right) + \left(1 - p^M \right) u\left(\left| \chi_j^E \left(z_j^E \right) - x_i^t \right| \right) - u\left(\left| \chi_i^t \left(z_i^t \right) - x_i^t \right| \right).$$

Following an election, an extreme type will betray the platform more severely and pay a higher cost of betrayal. However, at the same time, the ideal policy for an extreme type is further from the median policy than that of a moderate type, which means that his ideal policy is also further from the opponent's implemented policy. Thus, an extreme type has a higher disutility from the opponent's victory. Therefore, he finds it especially costly if the opponent wins, more so than a moderate type does. As a result, an extreme type has higher marginal benefit and cost than a moderate type. Since an extreme type has a higher marginal cost of betrayal, he does not have an incentive to choose a more moderate implemented policy than a moderate type (Lemma 3.4-(ii)). However, an extreme type has a higher marginal benefit, so he has an incentive to announce a more moderate platform than a moderate type (Lemma 3.4-(i)).

Therefore, when both candidates announce a pooling strategy, z_i^{M*}, an extreme type prefers his winning to the opponent winning. Thus, an extreme type does not have an incentive to lose. However, there is a possibility that an extreme type has an incentive to further approach the median policy and win with certainty, since he prefers his winning to the opponent winning at z_i^{M*}. To check this incentive of an extreme type, the off-path belief of voters must be specified.

3.3.3 Beliefs

Suppose that when a candidate deviates from z_i^{M*}, voters believe with a probability of one that the candidate is an extreme type. That is, I consider simple off-path beliefs as $p_i(M|z_i) = 0$. This simple off-path belief is *partially* based on the idea of the intuitive criterion in Cho and Kreps (1987). In a pooling equilibrium, a moderate type is indifferent to winning and losing at z_i^{M*}. Thus, a moderate type never chooses a more moderate platform than z_i^{M*} since her expected payoff decreases from the equilibrium expected-payoff even if voters believe this candidate to be a moderate type as a result of this deviation.[4] As a result, if a platform is announced that is more moderate than z_i^{M*}, this candidate must not be a moderate type. On the other hand, an extreme type has an incentive to announce a platform more extreme than his cut-off platform, given z_j^{M*} (i.e., $z_i^E\left(z_j^{M*}\right)$). Lemma 3.4 shows $z_i^E\left(z_j^{M*}\right)$ is more moderate than z_i^{M*}, so the off-path belief for a platform that lies between z_i^{M*} and $z_i^E\left(z_j^{M*}\right)$ should be $p_i(M|z_i) = 0$. The intuitive criterion cannot apply to any other off-path beliefs, so I simply assume that $p_i(M|z_i) = 0$ for all off-path strategies.

3.3.4 The existence of the pooling equilibrium

Voters do not know candidates' types in a pooling equilibrium, so their expected utility is the weighted average of the utility between a moderate and an extreme type. On the other hand, if an extreme type deviates by approaching the median policy, voters believe that this candidate's type is extreme because of the off-path belief. If an extreme type deviates to announce a sufficiently moderate platform, this extreme type can win over an uncertain opponent who chooses z_i^{M*}. Here, I denote z_i' such that it satisfies

$$-u\left(\left\|\chi_i^E\left(z_i'\right)-x_m\right\|\right)=-p^M u\left(\left\|\chi_j^M\left(z_j^{M*}\right)-x_m\right\|\right)$$
$$-\left(1-p^M\right)u\left(\left\|\chi_j^E\left(z_j^{M*}\right)-x_m\right\|\right) \tag{3.4}$$

The left-hand side is the utility of the median voter when extreme candidate i, who deviates to z_i' wins. The right-hand side is the expected utility of the median voter when candidate j, who announces the pooling platform z_j^{M*} wins. That is, the median voter is indifferent between candidates who announce z_i' and z_j^{M*}. If an extreme candidate announces a platform that is slightly more moderate than z_i', this candidate wins over an uncertain opponent. Figure 3.2(a) shows z_i' in the case of R when voters have linear utility. Note that because voters are uncertain about the type of the candidate who announces z_i^{M*}, an extreme type who deviates would need to implement a more moderate policy than an extreme type who chooses a pooling platform

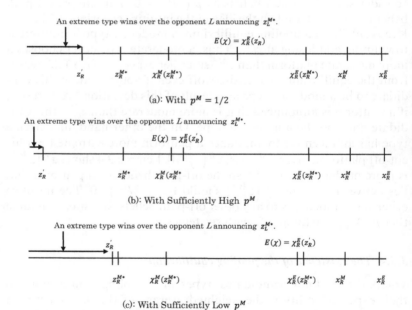

Figure 3.2 Pooling Equilibrium.
Let $E(x)=p^M\chi_R^M\left(z_R^{M*}\right)+\left(1-p^M\right)\chi_R^E\left(z_R^{M*}\right)$ denote the expected policy implemented by a candidate announcing z_R^{M*}. Suppose that voters have a linear utility. If an extreme type's platform is more moderate than z_R', such an extreme type wins over the opponent L announcing z_L^{M*}.

$\left(\chi_j^E\left(z_j^{M*}\right)\right)$ but does not need to implement a more moderate policy than a moderate type $\left(\chi_j^M\left(z_j^{M*}\right)\right)$, as shown in Figure 3.2(a).

An extreme type does not deviate by announcing a platform that is more moderate than z_i' if

$$-u\left(\left|\chi_i^E\left(z_i'\right)-x_i^E\right|\right)-c\left(\left|z_i'-\chi_i^E\left(z_i'\right)\right|\right)$$

$$\leq\frac{1}{2}\left[\begin{array}{l}-p^M u\left(\left|\chi_j^M\left(z_j^{M*}\right)-x_i^E\right|\right)-\left(1-p^M\right)u\left(\left|\chi_j^E\left(z_j^{M*}\right)-x_i^E\right|\right)\\-u\left(\left|\chi_j^E\left(z_j^{M*}\right)-x_i^E\right|\right)-c\left(\left|z_i^{M*}-\chi_i^E\left(z_i^{M*}\right)\right|\right)\end{array}\right] \quad (3.5)$$

The right-hand side is the extreme candidate i's expected utility when he stays in a pooling equilibrium. His expected utility from this deviation is slightly lower than the left-hand side. If (3.5) holds, this extreme type does not deviate, so a pooling equilibrium where all types announce z_i^{M*} exists. Now, denote

$$\overline{p}\equiv\frac{\left[\begin{array}{l}u\left(\left|\chi_i^E\left(z_j^{M*}\right)-x_i^E\right|\right)+u\left(\left|\chi_i^E\left(z_i^{M*}\right)-x_i^E\right|\right)+c\left(\left|z_i^{M*}-\chi_i^E\left(z_i^{M*}\right)\right|\right)\\-2\left[u\left(\left|\chi_i^E\left(z_i'\right)-x_i^E\right|\right)+c\left(\left|z_i'-\chi_i^E\left(z_i'\right)\right|\right)\right]\end{array}\right]}{u\left(\left|\chi_j^E\left(z_j^{M*}\right)-x_i^E\right|\right)-u\left(\left|\chi_j^M\left(z_j^{M*}\right)-x_i^E\right|\right)} \quad (3.6)$$

Then, condition (3.5) can be represented by $p^M\geq\overline{p}$, as shown in Proposition 3.3. This $\overline{p}$ is always positive and less than one.

Corollary 3.5

$\overline{p}\in(0,1)$.

   Proof: See Appendix 3.A.3.

   If $p^M\geq\overline{p}$, a pooling equilibrium exists. The next section shows that if $p^M<\overline{p}$, a semi-separating equilibrium exists. Thus, the previous corollary implies that for all parameter values and functional forms, if p^M is large enough, a pooling equilibrium always exists; otherwise, a semi-separating equilibrium exists.

   The intuitive reasoning is as follows. Suppose that L chooses z_L^{M*} as a pooling equilibrium, and R is an extreme type who originally announces z_R^{M*}. If p^M is high, extreme type R needs to announce a very moderate platform to win with certainty, because the expected utility to the median voter of choosing L is quite high given that there is a strong possibility that L is moderate and will implement a moderate policy. Thus, as in Figure 3.2(b), there is a significant distance between z_R' and R's ideal policy, so this deviation decreases his expected utility.

As a result, a pooling equilibrium exists. However, if p^M is sufficiently low, the expected utility of the median voter who chooses L is quite low. Thus, if R slightly approaches the median policy, the policy he implements improves for the median voter. That is, z'_R is closer to z_R^{M*}, as in Figure 3.2(c), so the extreme type will deviate from a pooling strategy.

Note that this pooling equilibrium exists in the broader value of the off-path beliefs. For example, suppose $p_i(M|z_i) = p^M$ if the platform is more extreme than z_i^{M*}. Then a candidate still has no incentive to deviate to a more extreme platform than z_i^{M*}, since he/she will then be certain to lose and the expected utility decreases or is unchanged by this deviation. A pooling equilibrium can exist when the off-path belief, $p_i(M|z_i)$, is lower than p^M for z_i, which is more extreme than z_i^{M*}.

3.4 Semi-separating equilibrium

If $p^M < \bar{p}$, a pooling equilibrium does not exist since an extreme type will deviate from z_i^{M*} to z'_i. In this case, a moderate type still announces one platform with certainty (a pure strategy). On the other hand, an extreme type chooses a mixed strategy: With some probability, an extreme type announces the same platform as a moderate type. With the remaining probability, an extreme type approaches the median policy. That is, if $p^M < \bar{p}$, a semi-separating equilibrium exists. I call an extreme type a "pooling extreme type" when he imitates the moderate type, but a "separating extreme type" when he approaches the median policy.

3.4.1 Definition and proposition

In a semi-separating equilibrium, a moderate type announces z_i^*, and an extreme type announces z_i^* with probability $\sigma^M \in (0,1)$, and $\bar{z}_i$ with the remaining probability, $1-\sigma^M$. I call this type of a semi-separating equilibrium a *two-policy semi-separating equilibrium*. This equilibrium is shown in Figure 3.3(a) and is defined as follows.

Definition 3.6

In a two-policy semi-separating equilibrium, a moderate type chooses z_i^*, and an extreme type chooses z_i^* with probability $\sigma^M \in (0,1)$, and $\bar{z}_i$ with probability $1-\sigma^M$.

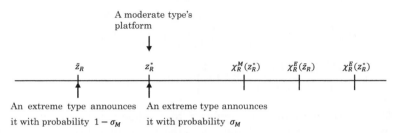

(a) A two-policy semi-separating equilibrium

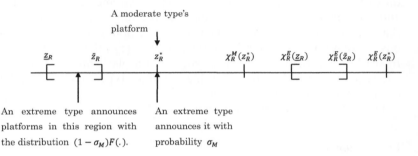

(b) A continuous semi-separating equilibrium

Figure 3.3 Semi-Separating Equilibrium.

Now, denote z_i^* (and z_j^*) such that it satisfies

$$-\frac{p^M}{p^M + \sigma^M \left(1 - p^M\right)} u\left(\left|\chi_j^M\left(z_j^*\right) - x_i^M\right|\right) - \frac{\sigma^M \left(1 - p^M\right)}{p^M + \sigma^M \left(1 - p^M\right)} \quad (3.7)$$
$$u\left(\left|\chi_j^E\left(z_j^*\right) - x_i^M\right|\right) = -u\left(\left|\chi_j^M\left(z_i^*\right) - x_i^M\right|\right) - c\left(\left|z_i^* - \chi_i^M\left(z_i^*\right)\right|\right)$$

where z_i^* and z_j^* are symmetric $\left(\left|x_m - z_i^*\right| = \left|x_m - z_j^*\right|\right)$. The left-hand side is the expected utility of moderate candidate i when the opponent promising z_j^* wins. The right-hand side is the utility of moderate candidate i when she wins. That is, a moderate type is indifferent between winning and losing and announces z_j^*.[5]

Next, denote $\bar{z}_i$ such that it satisfies

$$-u\left(\left|\chi_i^E(\bar{z}_i)-x_m\right|\right)>\frac{p^M}{p^M+\sigma^M\left(1-p^M\right)}u\left(\left|\chi_j^M\left(z_j^*\right)-x_m\right|\right)$$

$$-\frac{\sigma^M\left(1-p^M\right)}{p^M+\sigma^M\left(1-p^M\right)}u\left(\left|\chi_j^E\left(z_j^*\right)-x_m\right|\right)$$

(3.8)

That is, the median voter prefers $\bar{z}_i$ to z_j^* when $\bar{z}_i$ is announced by an extreme type. The right-hand side is the expected utility of the median voter when candidate j wins and announces z_j^*. The left-hand side is the expected utility of the median voter when extreme candidate i wins and announces $\bar{z}_i$. Moreover, $\bar{z}_i$ denotes the most extreme platform that satisfies (3.8). More precisely, because the policy space is continuous, there is no maximal (minimal) value of z_R (z_L) that satisfies (3.8). Instead, it is possible to define $\bar{z}_i$ such that a platform satisfies (3.8) with equality, and to assume that if an extreme type announces $\bar{z}_i$, he defeats an opponent who announces z_j^*.[6]

Then, the following proposition is derived.

Proposition 3.7

Suppose that the off-path beliefs of voters are $p_i(M|z_i)=0$. If $p^M<\bar{p}$, a semi-separating equilibrium exists in which a moderate type chooses z_i^*.

Proof: See Appendix 3.A.4.

3.4.2 Each player's choice

The details of the semi-separating equilibrium are given in Appendix 3.A.4; hence, I provide the intuitive reasoning in this section. Suppose again that when a candidate deviates from the equilibrium platform, voters believe with a probability of one that the candidate is an extreme type. That is, $p_i(M|z_i)=0$. This off-path belief is also partially based on the idea of the intuitive criterion. Since a moderate type is indifferent between winning losing $\left(\text{and announcing } z_j^*\right)$, a moderate type never chooses a platform more moderate than z_i^*. On the other hand, from Lemma 3.4, an extreme type prefers his winning to the winning of an opponent who announces z_j^* when he announces z_i^*.[7] Thus, an extreme type has an incentive to choose a more moderate platform $\left(\text{until his cut-off platform, given } z_j^*\right)$. The intuitive criterion cannot apply to any other off-path beliefs, so I simply assume that $p_i(M|z_i)=0$ for all off-path strategies.

The following summarizes each player's rational choice.

3.4.2.1 *Voters*

From the definition of $\bar{z}_i$ (Inequality (3.8)), an extreme type who announces $\bar{z}_i$ can win against an opponent who announces z_j^* (and tie with an opponent who announces $\bar{z}_j$). Suppose that candidate R (an extreme type) announces $\bar{z}_R$, while candidate L announces z_L^*. Voters can know that the type of R is extreme, but remain uncertain about the type of L who announces z_L^*, because an extreme type L will still pretend to be moderate and announce z_L^*. with probability σ^M. Therefore, to defeat L (i.e., to satisfy (3.8)), R does not need to implement a more moderate policy than a moderate type L. That is, $x_m - \chi_L^E\left(z_L^*\right) > \chi_R^E\left(\bar{z}_R\right) - x_m > x_m - \chi_L^M\left(z_L^*\right)$. In other words, for the median voter, a moderate type L will implement the best policy $\left(\chi_L^M\left(z_L^*\right)\right)$. However, if L wins, there is the possibility that L is an extreme type who implements the worst policy for the median voter $\left(\chi_L^E\left(z_L^*\right)\right)$. Thus, the median voter forgoes the chance of electing a moderate type L to avoid electing an extreme type L, and chooses the second-best candidate, R, who is a separating extreme type.

3.4.2.2 *A moderate type*

As a result of this off-path belief, a moderate type has no incentive to deviate from z_i^*. If a moderate type deviates to a more extreme platform than z_i^* or $z_i \in \left(\bar{z}_i, z_i^*\right)$, she will be certain to lose and her expected utility will remain unchanged as she is indifferent between her winning and the opponent winning at z_i^*. A moderate type has no incentive to approach the median policy by more than $\bar{z}_i$ since her utility from winning will then become lower than her utility when the opponent wins from Lemma 3.4.

3.4.2.3 *An extreme type*

An extreme type prefers his winning to the opponent winning and announcing z_j^* when he announces z_i^*. Thus, he does not have an incentive to deviate to a more extreme platform than z_i^* or $z_i \in \left(\bar{z}_i, z_i^*\right)$ since he will be certain to lose and his expected utility will decrease. When an extreme type announces $\bar{z}_i$, his disutility following a win and the cost of betrayal are higher, but the probability of winning is greater than when he announces z_i^*. Thus, an extreme type can be indifferent between $\bar{z}_i$ and z_i^* when $p^M < \bar{p}$.

3.4.3 The existence of the semi-separating equilibrium

A two-policy semi-separating equilibrium exists if $p^M < \bar{p}$ and

$$-u\left(\left|\chi_i^E\left(\bar{z}_i\right) - x_i^E\right|\right) - c\left(\left\|\bar{z}_i - \chi_i^E\left(\bar{z}_i\right)\right\|\right) \leq -u\left(\left\|\chi_j^E\left(\bar{z}_j\right) - x_i^E\right\|\right) \qquad (3.9)$$

When (3.9) holds, an extreme type has no incentive to defeat with certainty an extreme opponent who announces $\bar{z}_j$ by approaching the median policy. This is because (3.9) means that, for extreme candidate i, the utility when extreme opponent j who announces $\bar{z}_j$ wins is higher than the utility when i wins. However, an extreme type with $\bar{z}_i$ does not want to deviate to a more extreme platform because that would mean he would also lose to an opponent with z_j^* and his expected utility would decrease.

If (3.9) does not hold, an extreme type still has an incentive to converge by more than $\bar{z}_i$ to beat an extreme opponent who announces $\bar{z}_j$. Therefore, a two-policy semi-separating equilibrium does not exist, but a *continuous semi-separating equilibrium* does exist. In a continuous semi-separating equilibrium, an extreme type's mixed strategy includes z_i^* and a connected support, $[\bar{z}_L, \underline{z}_L]$ for L and $[\underline{z}_R, \bar{z}_R]$ for R, as shown in Figure 3.3(b). An extreme type chooses any platform in this support with probability $1 - \sigma^M$ and has a continuous distribution function, $F(.)$, within the support. More specifically, the distribution is $\left(1 - \sigma^M\right)F(.)$. Platform $\bar{z}_i$ is defined in the same way as $\bar{z}_i$ in a two-policy semi-separating equilibrium, so the basic results are the same as those of a two-policy semi-separating equilibrium. That is, a separating extreme type defeats an uncertain type (a moderate type and a pooling extreme type).

A semi-separating equilibrium exists in the broader value of the off-path beliefs. For example, suppose $p_i\left(M|z_i\right) = p^M / \left[p^M + \sigma^M\left(1 - p^M\right)\right]$ if the platform is more extreme than z_i^*. Then a candidate still has no incentive to deviate to a more extreme platform than z_i^*, since he/she will be certain to lose and the expected utility decreases or is unchanged by this deviation. Thus, a semi-separating equilibrium can exist when the off-path belief $p_i\left(M|z_i\right)$ is lower than $p^M / \left[p^M + \sigma^M\left(1 - p^M\right)\right]$ for $\bar{z}_i$, which is more extreme than z_i^*.

3.5 Discussions

3.5.1 Other equilibria

The previous sections only discuss a pooling equilibrium in which both types choose z_i^{M*} and a semi-separating equilibrium in which a moderate type chooses z_i^*. This section discusses other equilibria that may exist.

3.5.1.1 Separating equilibrium in which a moderate type wins

If a separating equilibrium exists in which a moderate type wins against an extreme type, both types must choose different platforms, and the moderate type must implement a more moderate policy than does the extreme type. However, such a separating equilibrium does not exist.

Proposition 3.8

There is no separating equilibrium in which a moderate type wins against an extreme type, regardless of off-path beliefs.

Proof: See Appendix 3.A.5.

This result is true because an extreme type always has an incentive to pretend to be moderate.

If a separating equilibrium exists in which a moderate type wins, the moderate type must prefer her winning to the opponent winning in this equilibrium. However, this means an extreme type also prefers to win over having the opponent win at the moderate type's platform, for the same reasoning given in Lemma 3.4. Moreover, if an extreme type deviates by pretending to be a moderate type, he will have a higher probability of winning. Thus, an extreme type has an incentive to deviate by choosing the moderate type's platform within this separating strategy.

This result contradicts that of Banks (1990) and Callander and Wilkie (2007) who show that there exists a separating equilibrium in which a moderate type wins.

3.5.1.2 Separating equilibrium in which an extreme type wins

A separating equilibrium in which an extreme type wins against a moderate type exists if off-path beliefs are $p_i(M|z_i) = 0$. For example, suppose that an extreme type announces a platform, $\hat{z}_i^E$, such that he is indifferent between winning and losing. To be precise, $\hat{z}_i^t$ satisfies $-u\left(\left|\chi_i^t\left(\hat{z}_i^t\right) - x_i^t\right|\right) - c\left(\left|\hat{z}_i^t - \chi_i^t\left(\hat{z}_i^t\right)\right|\right) = -u\left(\left|\chi_j^t\left(\hat{z}_j^t\right) - x_i^t\right|\right)$ where $\left|x_m - \hat{z}_i^t\right| = \left|x_m - \hat{z}_j^t\right|$. Suppose also that a moderate type announces $z_i^{M'}$, such that the moderate type will implement a more extreme policy than the extreme type. That is, $\left|x_m - \chi_i^E\left(\hat{z}_i^E\right)\right| < \left|x_m - \chi_i^M\left(z_i^{M'}\right)\right|$. As a result of the above off-path beliefs, although a moderate type approaches the median policy, voters believe that this candidate is an extreme type. To increase the probability of winning, a moderate type needs to approach the median policy in a significant way. This may decrease her expected utility. An extreme type does not deviate either, because he is indifferent between winning and losing. As a result, a separating equilibrium in which an extreme type wins exists.

However, this separating equilibrium is less important, since it exists in a very restricted set of off-path beliefs. For example, suppose that the off-path beliefs are $p_i(M|z_i) = p^M$ if the platform is more extreme than $\hat{z}_i^M$. Then a moderate type has an incentive to choose $\hat{z}_i^M$, where she is indifferent between winning and losing. If a moderate type announces $\hat{z}_i^M$, an extreme type does not have an incentive to win against a moderate type by committing to implement $\chi_i^E\left(z_i^E\right)$, which is more moderate than $\chi_i^M\left(\hat{z}_i^M\right)$, from Lemma 3.4. Therefore, this separating equilibrium does not exist with such off-path beliefs.

3.5.1.3   Other pooling and semi-separating equilibria

In the previous sections, I assume off-path beliefs as $p_i(M|z_i) = 0$. Under this assumption, there exist multiple pooling and semi-separating equilibria. First, there could be a pooling equilibrium in which both types announce a platform, say z_i^{M**}, that is more extreme than z_i^{M*} $\left(z_L^{M**} < z_i^{M*} \text{ and } z_R^{M*} < z_R^{M**}\right)$. Since the off-path belief is $p_i(M|z_i) = 0$, a moderate type needs to approach the median policy significantly to be sure of winning because voters believe that the candidate is extreme when he/she deviates from z_i^{M**}, regardless of the real type. Thus, a moderate type may not want to deviate. If an extreme type also has no incentive to deviate, a pooling equilibrium with z_i^{M**} exists. Second, there could be a semi-separating equilibrium in which a moderate type (and a pooling extreme type) announces a platform, say z_i^{**}, that is more extreme than z_i^*. A moderate type needs to approach the median policy significantly to win against an opponent who announces z_j^{**}, because this deviation leads voters to believe that the candidate is extreme. Therefore, a moderate type may have no incentive to deviate from z_i^{**}.

These equilibria (including a separating equilibrium in which an extreme type wins) have several problems. First, a moderate type does not want to approach the median policy because voters will likely misunderstand the candidate to be an extreme type. Second, if the off-path beliefs, $p_i(M|z_i)$, exceed zero for some off-path platforms, many of the above equilibria will be eliminated.[8] Thus, these equilibria exist for restricted values of off-path beliefs.

On the other hand, the equilibria analyzed in the previous sections exist in broader values of off-path beliefs than do other equilibria. As I discussed, a pooling equilibrium with z_i^{M*} exists when $p_i(M|z_i) = p^M$

if the platform is more extreme than z_i^{M*}. A semi-separating equilibrium with z_i^* exists when $p_i\left(M|z_i\right) = p^M / \left[p^M + \sigma^M \left(1 - p^M \right) \right]$ if the platform is more extreme than z_i^*.

3.5.2 Differences from past papers

3.5.2.1 Assumptions

As indicated in Chapter 1 (Section 1.3.2), there are two important differences between my model and those of Banks (1990) and Callander and Wilkie (2007): (1) candidates choose a policy to implement strategically and (2) they care about policy when they lose.

In a semi-separating equilibrium, an extreme type reveals his type by approaching the median policy with some probability since he can obtain a higher probability of winning. However, if candidates implement their own ideal policies automatically, voters only believe that an extreme type will implement his ideal policy, so a separating extreme type cannot increase his probability of winning by revealing his type. That is, a strategic choice of an implemented policy provides a *way* to win for an extreme type.

Suppose that a candidate does not care about policy when he loses. From (3.1), the expected utility of candidates is

$$\sum_{s=M,E}\left[p^s \int_{z_j^s} \pi_i^t\left(z_i^t, z_j^s\right) dF_j^s\left(z_j^s\right) \right]\left[-u\left(\left|\chi_i^t\left(z_i^t\right) - x_i^t\right|\right) - c\left(\left|z_i^t - \chi_i^t\left(z_i^t\right)\right|\right) \right]$$

Certainly, candidates will announce their ideal policy as platforms (and his/her expected utility is zero) because if not, they will incur disutility from the policy and a cost of betrayal, and the expected utility becomes negative. Thus, benefits from holding office should be introduced to induce candidates to approach the median policy. However, even with benefits from holding office, a semi-separating equilibrium does not exist when candidates do not care about policy after losing. An extreme type has a stronger incentive to prevent the opponent from winning in my model but to a lesser extent when they do not care about an opponent's policy. Thus, caring about policy after losing provides an extreme type with an *incentive* to win.

3.5.2.2 Universal divinity

Banks (1990), Callander and Wilkie (2007), and Huang (2010) employ universal divinity, as introduced by Banks and Sobel (1987).

If universal divinity is applied to my model, in short, as Lemma 3.4 shows, an extreme type always has a greater incentive to announce a more moderate *platform* than does a moderate type. This means that a moderate type always has a stronger incentive to announce a more extreme platform than does an extreme type. Suppose a pooling equilibrium at z_i^{M*}. With universal divinity, if a platform is more moderate than z_i^{M*}, $p_i(M|z_i) = 0$. If not, $p_i(M|z_i) = 1$. With these off-path beliefs, both have an incentive to deviate to a more extreme platform than z_i^{M*}, be thought a moderate type by voters, and win. For the same reasons, a semi-separating equilibrium does not exist, and only a separating equilibrium exists in which an extreme type wins against a moderate type, which is discussed in Section 3.5.1. However, such a separating equilibrium seems peculiar, as I discussed. Moreover, in this separating equilibrium, an extreme type always wins against a moderate type, so my main result does not change.[9]

3.6 Applications: Turkey, Japan, and the UK

The important equilibrium is a semi-separating equilibrium. Thus, I discuss the national elections in Turkey, Japan, and the UK as examples of this equilibrium.[10] Since my model is simple, these examples do not exactly match my model. For example, my model considers only two symmetric parties, a plurality voting system, and a single-policy issue. In Japan, Turkey, and the UK, there are more than two parties. Turkey uses proportional representation (PR), and Japan employs parallel voting, including both PR and single-member districts. Additionally, it is rare to have an election with a single-policy issue and symmetric parties.[11]

However, I show here that these examples have at least the following four important characteristics of a semi-separating equilibrium: (1) one party's platform is more moderate than the other party's platform; (2) voters guess that such a party is an extreme type; (3) voters are uncertain whether the opposition(s) is an extreme type; and (4) the party that announces the more moderate platform wins the election.

3.6.1 Turkey

In Turkish politics, there are two large groups, namely, Political Islam and secular parties. Broadly speaking, secularists, represented by parties such as the Republican People's Party, support democratic systems and politico-religious separation. Political Islam, represented by the Justice and Development Party (AKP), wants to introduce Islamic doctrines into some policies. The AKP and the prime minister, Recep Erdogan, have promoted the AKP as the party of reform, a party that supports democratic systems, including politico-religious separation. Most citizens support secularism in Turkey, and the AKP's promises were almost the same as those encapsulated in the opponent's policies. Nevertheless, voters realized that the AKP is the extreme Islamic party (Dağt, 2006). On the other hand, in the 2007 Turkey presidential election, the Turkish military, which supports secularism, stated that "the Turkish armed forces have been monitoring the situation with concern." People interpreted this as a threat of a coup and started to worry that the secular parties would support extreme secular policies, such as using violence against Political Islam. Thus, voters were uncertain about the secular party's type. Finally, the AKP won the 2007 elections.

Therefore, this case has the four characteristics of a semi-separating equilibrium: (1) the AKP compromised greatly by promising politico-religious separation; (2) people recognized the AKP as an extreme party that still supported Political Islam; (3) the type of the opponents became uncertain for voters after the threat of a coup; and (4) the AKP won.

3.6.2 *Japan*

In Japan between 1998 and 2016, there were two main parties, the Liberal Democratic Party (LDP), which supports increasing government spending on, for example, public works to sustain rural areas, and the Democratic Party of Japan (DPJ), which supports economic reforms and reducing government debt. In 2001, the LDP chose Junichiro Koizumi as their leader. Koizumi promised to implement economic reforms such as reducing government works and debt, and moreover, promised to "destroy" the (traditional) LDP. After the "great depression" during the 1990s, many Japanese supported implementing economic reforms rather than traditional economic policies, so the LDP's position should have been further from the median policy than the DPJ after the 1990s. Indeed, voters were afraid that the LDP would not implement the economic reforms (Mulgan, 2002). The opposition, the DPJ, had no experience in government, so voters remained uncertain about the party. Finally, the LDP led by Koizumi won the elections in 2001, 2003, and 2005.

This case also satisfies the four characteristics of a semi-separating equilibrium: (1) the LDP compromised greatly by promising to destroy the traditional LDP; (2) people still believed that the LDP was too conservative; (3) the DPJ's type was uncertain for voters, as it had no experience in government; and (4) the LDP won.

3.6.3   The U.K.

In the UK, there are two major parties, the Conservative Party and the Labour Party. Broadly speaking, while the Conservative Party supports the free market, the Labour Party is famous for its support of socialist policies and for being supported by labor unions. The Labour Party had not been in government since 1979 because many citizens did not support socialist policies. In 1994, the Labour Party chose Tony Blair as their leader, and he promised the "Third Way" and free-market policies. Most members of the Labour Party supported Blair, although some, such as members of labor unions, still supported socialist policies. This means that the Labour Party compromised greatly by choosing Blair as their leader. On the other hand, voters were uncertain about the Conservative Party preferences because of infighting between factions. As a result, the Labour Party won the 1997 election (Clarke, 2004).

This case matches the four characteristics of a semi-separating equilibrium, as follows: (1) the Labour Party compromised greatly by promising the Third Way; (2) people believed that the Labour Party supported socialist policies, since many members still supported these policies; (3) the type of the Conservative Party was uncertain for voters because of intra-party conflict; and (4) the Labour Party won.

3.7 Summary

This chapter examines how an extreme candidate wins against a moderate candidate to provide one reason to have political polarization, based on the model of partially binding platforms with asymmetric information. There are two main equilibria, namely, semi-separating and pooling, and voters cannot determine a candidate's political preferences in any equilibrium. In a pooling equilibrium, an extreme candidate imitates a moderate candidate by announcing the same platform as the moderate candidate. In a semi-separating equilibrium, an extreme candidate imitates a moderate candidate with some probability, and with the remaining probability reveals his own preferences by approaching the median policy. An extreme candidate who reveals his preference type will defeat an unknown candidate who may be moderate or extreme, even though this extreme candidate will implement a more extreme policy than the moderate candidate. This is because voters wish to avoid electing an extreme type who imitates a moderate candidate and will implement the most extreme policy. As a result, an extreme candidate has a higher expected probability of winning than a moderate candidate. The important reason for this result is that an extreme candidate has a stronger incentive to prevent an opponent from winning because his ideal policy is further from the opponent's policy than is the ideal policy of a moderate candidate.

Combining the implications from Chapters 2 and 3, the following result is obtained. When voters know little about candidates and are uncertain about their policy preferences, an extreme candidate tends to win an election. By contrast, when voters have sufficient information about candidates, a more moderate candidate tends to win. Therefore, political polarization tends to occur when voters have insufficient information about candidates.

3.A Appendix: proofs

3.A.1 Proposition 3.3

First, a candidate has no incentive to deviate to a more extreme platform than z_i^{M*} or $z_i \in \left[z_i', z_i^{M*} \right)$ because off-path beliefs are $p_i(M|z_i) = 0$. Hence, this candidate will be certain to lose and the expected utility will decrease for an extreme type (from Lemma 3.4) and will remain unchanged for a moderate type (from the definition of z_i^{M*}). A moderate type does not have an incentive to approach the median policy by more than z_i', because her utility from winning will become lower than her utility if her opponent wins (from the definition of z_i^{M*}). An extreme type also does not have an incentive to deviate by approaching the median policy by more than z_i' when $p^M \geq \bar{p}$, from the discussion in Subsection 3.4.2. As a result, this is an equilibrium in which both candidates announce z_i^{M*}.

∎

3.A.2 Lemma 3.4

Consider a case of R, without loss of generality. Let $\chi_R^t = \chi_R^t \left(z_R^t(z_L) \right)$ denote the situation where the utility when type t R wins is the same as the expected utility for type t R when L wins, given z_L. This means that

$$
\begin{aligned}
& p^M u \left(x_R^t - \chi_L^M \right) + \left(1 - p^M \right) u \left(x_R^t - \chi_L^E \right) - \\
& u \left(x_R^t - \chi_R^t \right) = c \left(\chi_R^t - z_R^t \left(\chi_R^t \right) \right)
\end{aligned}
\tag{3.10}
$$

where $z_R^t\left(\chi_R^t\right)$ represents the platform in which the candidate implements χ_R^t. First, I prove the second statement.

(ii) Differentiate both sides of (3.10) with respect to x_R^t, given the opponent's strategies $\left(\chi_L^M = \chi_L^M\left(z_L\right)\right.$ and $\left.\chi_L^E = \chi_L^E\left(z_L\right)\right)$. Then,

$$
\begin{aligned}
&p^M u'\left(x_R^t - \chi_L^M\right) + \left(1 - p^M\right)u'\left(x_R^t - \chi_L^E\right) - u'\left(x_R^t - \chi_R^t\right)\left(1 - \frac{\partial \chi_R^t}{\partial x_R^t}\right) \\
&= c'\left(x_R^t - z_R^t\left(\chi_R^t\right)\right)\left[\frac{\partial \chi_R^t}{\partial x_R^t} - \left(\frac{\partial z_R^t\left(\chi_R^t\right)}{\partial x_R^t} + \frac{\partial z_R^t\left(\chi_R^t\right)}{\partial x_R^t}\frac{\partial \chi_R^t}{\partial x_R^t}\right)\right]
\end{aligned}
\tag{3.11}
$$

From Lemma 3.1, $u'\left(x_R^t - \chi_R^t\right) = c'\left(\chi_R^t - z_R^t\left(\chi_R^t\right)\right)$. Moreover, fix χ_R^t and differentiate $u'\left(x_R^t - \chi_R^t\right) = c'\left(\chi_R^t - z_R^t\left(\chi_R^t\right)\right)$ with respect to x_R^t. Then,

$$
\frac{\partial z_R^t\left(\chi_R^t\right)}{\partial x_R^t} = -\frac{u''\left(x_R^t - \chi_R^t\right)}{c''\left(\chi_R^t - z_R^t\left(\chi_R^t\right)\right)} < 0.
$$

Substitute these values into (3.11), which then becomes

$$
\frac{\partial \chi_R^t}{\partial x_R^t} = \frac{\dfrac{u''\left(x_R^t - \chi_R^t\right)c'\left(\chi_R^t - z_R^t\left(\chi_R^t\right)\right)}{c''\left(\chi_R^t - z_R^t\left(\chi_R^t\right)\right)} - \left[\begin{array}{l}p^M u'\left(x_R^t - \chi_L^M\right) + \left(1 - p^M\right) \\ u'\left(x_R^t - \chi_L^E\right) - u\left(x_R^t - \chi_R^t\right)\end{array}\right]}{u'\left(x_R^t - \chi_R^M\right)\dfrac{\partial z_R^t\left(\chi_R^t\right)}{\partial x_R^t}}
\tag{3.12}
$$

If (3.12) is positive, an extreme type will implement a more extreme policy than will a moderate type. In the same way as $\partial z_R^t\left(\chi_R^t\right)/\partial \chi_R^t$ was derived,

$$
\frac{\partial z_R^t\left(\chi_R^t\right)}{\partial \chi_R^t} = 1 + \frac{u''\left(x_R^t - \chi_R^t\right)}{c''\left(\chi_R^t - z_R^t\left(\chi_R^t\right)\right)} > 0,
$$

so the denominator of (3.12) is positive. To prove that (3.12) is positive, it is sufficient to show that the numerator of (3.12) is positive. In other words,

$$
\begin{aligned}
&\frac{p^M u'\left(x_R^t - \chi_L^M\right) + \left(1 - p^M\right)u'\left(x_R^t - \chi_L^E\right) - u'\left(x_R^t - \chi_R^t\right)}{u''\left(x_R^t - \chi_R^t\right)} \\
&< \frac{c'\left(x_R^t - z_R^t\left(\chi_R^t\right)\right)}{c''\left(\chi_R^t - z_R^t\left(\chi_R^t\right)\right)}
\end{aligned}
\tag{3.13}
$$

Note that, from (3.10) and Lemma 3.1,

$$
\frac{p^M u\left(x_R^t - \chi_L^M\right) + \left(1 - p^M\right) u\left(x_R^t - \chi_L^E\right) - u\left(x_R^t - \chi_R^t\right)}{u'\left(x_R^t - \chi_R^t\right)}
$$

$$
= \frac{c\left(x_R^t - z_R^t\left(\chi_R^t\right)\right)}{c'\left(\chi_R^t - z_R^t\left(\chi_R^t\right)\right)} \tag{3.14}
$$

Since I assume that $c'(d)/c(d)$ strictly decreases as d increases (Assumption 1), $c'\left(\chi_R^t - z_R^t\left(\chi_R^t\right)\right)/c''\left(\chi_R^t - z_R^t\left(\chi_R^t\right)\right) > c\left(\chi_R^t - z_R^t\left(\chi_R^t\right)\right)/c'\left(\chi_R^t - z_R^t\left(\chi_R^t\right)\right)$. The right-hand side of (3.13) is greater than the right-hand side of (3.14). Therefore, if the left-hand side of (3.13) is less than the left-hand side of (3.14), (3.13) holds. This means

$$
p^M \left(\frac{u'\left(x_R^t - \chi_L^M\right)}{u''\left(x_R^t - \chi_R^t\right)} - \frac{u'\left(x_R^t - \chi_L^M\right)}{u'\left(x_R^t - \chi_R^t\right)} \right) +
$$

$$
\left(1 - p^M\right) \left(\frac{u'\left(x_R^t - \chi_L^E\right)}{u''\left(x_R^t - \chi_R^t\right)} - \frac{u\left(x_R^t - \chi_L^E\right)}{u'\left(x_R^t - \chi_R^t\right)} \right)
$$

$$
< \frac{u'\left(x_R^t - \chi_R^t\right)}{u''\left(x_R^t - \chi_R^t\right)} - \frac{u\left(x_R^t - \chi_R^t\right)}{u'\left(x_R^t - \chi_R^t\right)}
$$

Since I assume that $u'(d)/u(d)$ strictly decreases as d increases (Assumption 3), the right-hand side is positive. If $\chi_L^E = \chi_L^M = \chi_R^t$, both sides are the same. If χ_L^E and χ_L^M become further from x_R^t than χ_R^t, the left-hand side decreases. The reason is as follows. I differentiate $u'\left(x_R^t - \chi_L^k\right)/u''\left(x_R^t - \chi_R^t\right) - u\left(x_R^t - \chi_L^k\right)/u'\left(x_R^t - \chi_R^t\right)$ with respect to $x_R^t - \chi_L^k$, then $u''\left(x_R^t - \chi_L^k\right)/u''\left(x_R^t - \chi_R^t\right) - u'\left(x_R^t - \chi_L^k\right)/u'\left(x_R^t - \chi_R^t\right)$. This value is negative because I assume that $u''(d)/u'(d)$ strictly decreases as d increases. As a result, the left-hand side of (3.13) is less than the left-hand side of (3.14), so (3.13) holds, and (3.12) is positive.

(i) The first statement is true because

$$
\frac{\partial z_R^t\left(\chi_R^t\right)}{\partial x_R^t} + \frac{\partial z_R^t\left(\chi_R^t\right)}{\partial \chi_R^t}\frac{\partial \chi_R^t}{\partial x_R^t}
$$

$$
= -\frac{1}{u'\left(x_R^t - \chi_R^t\right)}\begin{bmatrix} p^M u'\left(x_R^t - \chi_L^M\right) + \left(1 - p^M\right) u'\left(x_R^t - \chi_L^E\right) \\ -u'\left(x_R^t - \chi_R^t\right) \end{bmatrix} < 0.
$$

∎

3.A.3 Corollary 3.5

When $p^M = 0$, $z_i' = z_i^{M*}$, from (3.4). Thus, the left-hand side of (3.5) is $-u\left(\left\|\chi_i^E\left(z_i^{M*}\right)-x_i^E\right\|\right)-c\left(\left\|z_i^{M*}-\chi_i^E\left(z_i^{M*}\right)\right\|\right)$ and is strictly greater than the right-hand side, since $-u\left(\left\|\chi_j^E\left(z_j^{M*}\right)-x_i^E\right\|\right)<-u\left(\left\|\chi_i^E\left(z_i^{M*}\right)-x_i^E\right\|\right)$ $-c\left(\left\|z_i^{M*}-\chi_i^E\left(z_i^{M*}\right)\right\|\right)$. That is, (3.5) does not hold when $p^M = 0$.

From (3.4), when $p^M = 1$, $z_i' = z_i^E\left(\chi_i^M\left(z_i^{M*}\right)\right)$. Here, if an extreme type announces the platform $z_i^E\left(\chi_i^M\left(z_i^{M*}\right)\right)$, he will implement the policy of the moderate type, $\chi_i^M\left(z_i^{M*}\right)$. The left-hand side of (3.5) is $-u\left(\left\|\chi_i^M\left(z_i^{M*}\right)-x_i^E\right\|\right)-c\left(\left\|z_i^E\left(\chi_i^M\left(z_i^{M*}\right)\right)-\chi_i^M\left(z_i^{M*}\right)\right\|\right)$, which is strictly less than its right-hand side, since $-u\left(\left\|\chi_j^M\left(z_j^{M*}\right)-x_i^E\right\|\right)>-u\left(\left\|\chi_i^M\left(z_i^{M*}\right)-x_i^E\right\|\right)-c\left(\left\|z_i^E\left(\chi_i^M\left(z_i^{M*}\right)\right)-\chi_i^M\left(z_i^{M*}\right)\right\|\right)$, from Lemma 3.4, and $-u\left(\left\|\chi_i^E\left(z_i^{M*}\right)-x_i^E\right\|\right)-c\left(\left\|z_i^{M*}-\chi_i^E\left(z_i^{M*}\right)\right\|\right)>-u\left(\left\|\chi_i^M\left(z_i^{M*}\right)-x_i^E\right\|\right)-c\left(\left\|z_i^E\left(\chi_i^M\left(z_i^{M*}\right)\right)-\chi_i^M\left(z_i^{M*}\right)\right\|\right)$. That is, (3.5) is satisfied when $p^M = 1$.

Both sides of (3.5) change continuously with p^M, so there always exists $p^M = \overline{p} \in (0,1)$ in which both sides of (3.5) are equal.

∎

3.A.4 Proposition 3.7

The precise definition of a semi-separating equilibrium is as follows. Denote the expected utility of an extreme type who announces z_i as $V_i^E(z_i)$.

Definition 3.9 *A continuous semi-separating equilibrium is a collection* $\left(z_i^*,\sigma^M,F(.),\Pi\right)$ *and a two-policy semi-separating equilibrium is a collection* $\left(z_i^*,\sigma^M,\overline{z}_i,\Pi\right)$, *where* z_i^*, *is a platform chosen by a moderate type,* σ^M *is the probability of an extreme type choosing* z_i^* *in a mixed strategy,* $F(.)$ *is a distribution function with the support of* $[\overline{z}_L,\underline{z}_L]$ *for L and* $[\underline{z}_R,\overline{z}_R]$ *for R, and* Π *is a scalar, such that: (a)* $\Pi=V_i^E(z_i)=V_i^E\left(z_i^*\right)$, *for all* z_i *in support of* $F(.)$ *in a continuous semi-separating equilibrium; and (b)* $\Pi=V_i^E\left(z_i^*\right)=V_i^E(\overline{z}_i)$ *in a two-policy semi-separating equilibrium.*

Define σ^M and Π

First, I discuss a continuous semi-separating equilibrium.[12] When an extreme type announces z_i^*, the expected utility is

$$V_i^E\left(z_i^*\right)=\left(\frac{1}{2}\right)\left[-p^M u\left(\left|\chi_j^M\left(z_j^*\right)-x_i^E\right|\right)-\sigma^M\left(1-p^M\right)u\left(\left|\chi_j^E\left(z_j^*\right)-x_i^E\right|\right)-\left(p^M+\sigma^M\left(1-p^M\right)\right)\left\{u\left(\left|\chi_i^E\left(z_i^*\right)-x_i^E\right|\right)+c\left(\left|z_i^*-\chi_i^E\left(z_i^*\right)\right|\right)\right\}\right]-\left(1-\sigma^M\right)\left(1-p^M\right)\int_{z_i}^{\bar{z}_i}u\left(\left|\chi_j^E\left(z_j\right)-x_i^E\right|\right)dF\left(z_j\right).$$

When an extreme type announces $\bar{z}_i$, the expected utility is $V_i^E\left(\bar{z}_i\right)=\left(p^M+\sigma^M\left(1-p^M\right)\right)\left[-u\left(\left|\chi_i^E\left(\bar{z}_i\right)-x_i^E\right|\right)-c\left(\left|\bar{z}_i-\chi_i^E\left(\bar{z}_i\right)\right|\right)\right]-\left(1-\sigma^M\right)\left(1-p^M\right)\int_{z_i}^{\bar{z}_i}u\left(\left|\chi_j^E\left(z_j\right)-x_i^E\right|\right)dF\left(z_j\right).$ The

value of σ^M is decided at the point at which the extreme type's expected utilities under z_i^* and $\bar{z}_i$ are the same:

$$\frac{1}{2}\left[\begin{array}{c}-\dfrac{p^M}{p^M+\sigma^M\left(1-p^M\right)}u\left(\left|\chi_j^M\left(z_j^*\right)-x_i^E\right|\right)\\[2mm]-\dfrac{\sigma^M\left(1-p^M\right)}{p^M+\sigma^M\left(1-p^M\right)}u\left(\left|\chi_j^E\left(z_j^*\right)-x_i^E\right|\right)\\[2mm]-u\left(\left|\chi_i^E\left(z_i^*\right)-x_i^E\right|\right)-c\left(\left|z_i^*-\chi_i^E\left(z_i^*\right)\right|\right)\end{array}\right] \tag{3.15}$$
$$=-u\left(\left|\chi_i^E\left(\bar{z}_i\right)-x_i^E\right|\right)-c\left(\left|\bar{z}_i-\chi_i^E\left(\bar{z}_i\right)\right|\right)$$

When $\sigma^M=1$, the left-hand side of (3.15) is less than the right-hand side because (3.5) does not hold. If the left-hand side is greater than the right-hand side when $\sigma^M=0$, the value of $\sigma^M\in(0,1)$ under which an extreme type is indifferent between z_i^* and $\bar{z}_i$ exists. The following condition means that the left-hand side is greater than the right-hand side of (3.15) when $\sigma^M=0$:

$$-\frac{1}{2}\left[u\left(\left|\chi_j^M\left(z_j^*\right)-x_i^E\right|\right)-u\left(\left|\chi_i^E\left(z_i^*\right)-x_i^E\right|\right)+c\left(\left|z_i^*-x_i^E\left(z_i^*\right)\right|\right)\right] \tag{3.16}$$
$$>-u\left(\left|\chi_i^E\left(\bar{z}_i\right)-x_i^E\right|\right)-c\left(\left|\bar{z}_i-\chi_i^E\left(\bar{z}_i\right)\right|\right)$$

First, $-u\left(\left|\chi_i^E\left(z_i^*\right)-x_i^E\right|\right)-c\left(\left|z_i^*-x_i^E\left(z_i^*\right)\right|\right)>-u\left(\left|\chi_i^E\left(\bar{z}_i\right)-x_i^E\right|\right)-c\left(\left|\bar{z}_i-x_i^E\left(\bar{z}_i\right)\right|\right)$ because $\bar{z}_i$ is more moderate than z_i^*. Second, $\bar{z}_i$ is the platform with which an extreme type can defeat a moderate type who announces z_i^*. When σ^M becomes zero, voters guess that a candidate announcing z_i^* is a moderate type. From the definition of $\bar{z}_i$, an extreme type's implemented policy, $\chi_i^E(\bar{z}_i)$, needs to be more moderate than a moderate type's implemented policy, $\chi_i^M(z_i^*)$. From Lemma 3.4, a moderate type has a greater incentive to choose a more moderate *implemented policy* than an extreme type, and a moderate type is indifferent to winning or losing at z_i^*. This means that $-u\left(\left|\chi_j^M\left(z_j^*\right)-x_i^E\right|\right)>-u\left(\left|\chi_i^E\left(\bar{z}_i\right)-x_i^E\right|\right)-c\left(\left|\bar{z}_i-\chi_i^E\left(\bar{z}_i\right)\right|\right)$. As a result, (3.16) holds, so a value of $\sigma^M\in(0,1)$ under which an extreme type is indifferent between $\bar{z}_i$ and z_i^* exists.

The other bound of support for $F(.)$

The distribution function, $F(.)$, satisfies the following lemma.

Lemma 3.10

Suppose that a continuous semi-separating equilibrium exists. In such an equilibrium, $F(.)$ is continuous with connected support.

Proof: If $F(.)$ has a discontinuity at some policy, say z_i' (i.e., $F(z_i'+)>F(z_i'-)$) there is a strictly positive probability that an opponent also chooses z_j' (the probability density function is $f(z_j')>0$). If this candidate approaches the median policy by an infinitesimal degree, it increases the probability of winning by $f(z_j')/2>0$. On the other hand, because this approach is minor, the expected utility changes by slightly less than $(1/2)f(z_j')\left[-u\left(\left|\chi_i^E(z_i')-x_i\right|\right)-c\left(\left|z_i'-\chi_i^E(z_i')\right|\right)-\left\{-u\left(\left|\chi_j^E(z_j')-x_i\right|\right)\right\}\right]$ and is positive (or negative). This implies that if $F(.)$ has a discontinuity, it cannot be part of a continuous semi-separating equilibrium. Assume that $F(.)$ is constant in some region, $[z_1,z_2]$, in the convex hull of the support. If a candidate chooses z_1, he has an incentive to deviate to z_2 because the probability of winning does not change. However, the implemented policy will approach the candidate's own ideal policy, so the expected utility increases. Thus, the support of $F(.)$ must be connected.

∎

At $\underline{z}_i$, the expected utility is $V_i^E(\underline{z}_i)=-u\left(\left|\chi_i^E(\underline{z}_i)-x_i^E\right|\right)+c\left(\left|\underline{z}_i-\chi_i^E(\underline{z}_i)\right|\right)$ because, from Lemma 3.10, $F(\underline{z}_L)=0$, and the probability of winning is one. If (3.9) does not hold, $V_i^E(\underline{z}_i)$ is higher than $V_i^E(\bar{z}_i)$ when $\underline{z}_i=\bar{z}_i$, so $\underline{z}_i\neq\bar{z}_i$ in equilibrium. Therefore, a continuous semi-separating equilibrium exists. If (3.9) holds, the extreme bound and the moderate bound are equivalent (a two-policy semi-separating equilibrium). Suppose that (3.9) does not hold. In equilibrium, $V_i^E(\underline{z}_i)$ and $V_i^E(\bar{z}_i)$ should be the same, so $\underline{z}_i$ and $F(.)$ should satisfy the following equation. Further, suppose R, without loss of generality, then

$$-u\left(x_R^E-\chi_R^E\left(\underline{z}_R\right)\right)-c\left(\chi_R^E\left(\underline{z}_R\right)-\underline{z}_R\right)$$
$$=\left(p^M+\sigma^M\left(1-p^M\right)\right)\left[-u\left(x_R^E-\chi_R^E\left(\bar{z}_R\right)\right)-c\left(\chi_R^E\left(\bar{z}_R\right)-\bar{z}_R\right)\right]$$
$$-\left(1-\sigma^M\right)\left(1-p^M\right)\int_{\bar{z}_L}^{zL}u\left(x_R^E-\chi_L^E\left(z_L\right)\right)dF\left(z_L\right) \tag{3.17}$$

I assume that the two candidates' positions are symmetric, so when $\underline{z}_R$ decreases, $\underline{z}_L$ increases. Then, $V_R^E(\bar{z}_R)$ increases because $\int_{\bar{z}_L}^{zL}u\left(x_R^E-\chi_L^E\left(z_L\right)\right)dF\left(z_L\right)$ decreases while $V_R^E(\underline{z}_R)$ decreases. In addition, $F(.)$ adjusts the value of $\int_{\bar{z}_L}^{zL}u\left(x_R^E-\chi_L^E\left(z_L\right)\right)dF\left(z_L\right)$. Thus, there exist combinations of $\bar{z}_i$ and $F(.)$ that satisfy (3.17).

I denote $\hat{z}_i^E$ such that $-u\left(\left\|\chi_j^E\left(\hat{z}_j^E\right)-x_i^E\right\|\right)=-u\left(\left\|\chi_i^E\left(\hat{z}_i^E\right)-x_i^E\right\|\right)-c\left(\left\|\hat{z}_i^E-\chi_i^E\left(\hat{z}_i^E\right)\right\|\right)$. The moderate bound, $\underline{z}_i$, should be more extreme than $\hat{z}_i^E$. If $\underline{z}_i$ is more moderate than $\hat{z}_i^E$, it means $-u\left(\left\|\chi_j^E\left(\underline{z}_j\right)-x_i^E\right\|\right)>-u\left(\left\|\chi_i^E\left(\underline{z}_i\right)-x_i^E\right\|\right)-c\left(\left\|\underline{z}_i-\chi_i^E\left(\underline{z}_i\right)\right\|\right)$. Thus, an extreme type with $\underline{z}_i$ has an incentive to deviate and lose to an extreme opponent with a platform close to $\underline{z}_j$. Any platform in the support of $F(.)$, say z_i', needs to satisfy $-u\left(\left\|\chi_j^E\left(z_j'\right)-x_i^E\right\|\right)>-u\left(\left\|\chi_i^E\left(z_i'\right)-x_i^E\right\|\right)-c\left(\left\|z_i'-\chi_i^E\left(z_i'\right)\right\|\right)$ to avoid deviating to lose. Therefore, $\chi_i^E(\underline{z}_i)$ is more extreme than $\chi_i^M\left(z_i^*\right)$, because $\chi_i^E\left(\hat{z}_i^E\right)$ is more extreme than $\chi_i^M\left(z_i^*\right)$.

Define $F(.)$

Suppose R, without loss of generality. Let

$$X(z_L')=\int_{zL}^{zL}u\left(x_R^E-\chi_L^E\left(z_L\right)\right)dF\left(z_L\right).$$

For any $z_R'\in\left(\underline{z}_R,\bar{z}_R\right)$, the expected utility should be the same as Π.[13] This means that

$$F_X\left(z_R'\right)=\frac{\Pi+u\left(x_R^E-\chi_R^E\left(z_R'\right)\right)+c\left(x_R^E\left(z_R'\right)-z_R'\right)X\left(z_L'\right)}{\left(1-\sigma^M\right)\left(1-p^M\right)u\left(x_R^E-\chi_R^E\left(z_R'\right)\right)+c\left(\chi_R^E\left(z_R'\right)-z_R'\right)}$$

The distribution function, $F_X(.)$, is defined by the above equation for any platform in support of $F(.)$, given $X(z_L')$. When $F_X\left(z_R'\right)=0$, it is

$\Pi + u\left(x_R^E - \chi_R^E\left(z_R'\right)\right) + c\left(\chi_R^E\left(z_R'\right) - z_R'\right)X\left(z_L'\right) = 0$. This equation holds if and only if $z_R' = \underline{z}_R$ and $X\left(z_L'\right) = 0$, such that $\Pi = V_R^E\left(\underline{z}_R\right)$. Then, $X\left(z_L'\right) = 0$ if and only if $z_L' = \underline{z}_L$. Therefore, when z_R' and z_L' become $\underline{z}_R$ and $\underline{z}_L$, respectively, $F\left(z_R'\right)$ becomes zero.

When $F\left(z_R'\right) = 1$, it is $\Pi = \left(p^M + \sigma^M\left(1 - p^M\right)\right)\left[-u\left(x_R^E - \chi_R^E\left(z_R'\right)\right) - c\left(\chi_R^E\left(z_R'\right) - z_R'\right)\left(1 - p^M\right)X\left(z_L'\right)\right]$. This equation holds if and only if $z_R' = \overline{z}_R$ and $X\left(z_L'\right) = \int_{\overline{z}_L}^{\underline{z}_L} u\left(x_R^E - \chi_L^E\left(z_L\right)\right)dF\left(z_L\right)$, such that $\Pi = V_R^E\left(\overline{z}_R\right)$. This means that when z_R' and z_L' become $\overline{z}_R$ and $\overline{z}_L$, respectively, $F\left(z_R'\right)$ becomes one.

When z_L' satisfies $|x_m - z_L'| = |x_m - z_R'|$, that is, $F(.)$ is symmetric for both candidates, the value of $X\left(z_L'\right)$ increases continuously as $z_R'\left(z_L'\right)$ becomes more extreme. Therefore, if the platform moves from $\underline{z}_R$ to $\overline{z}_R$, $F\left(z_R'\right)$ increases from zero to one. Thus, if $F(.)$ is symmetric for both candidates, $F_i(.)$ can be defined for $i = L, R$.

An extreme type does not deviate

An extreme type does not deviate to a more moderate platform than $\underline{z}_i$, as the probability of winning is still one. However, the cost of betrayal and the disutility following a win will increase.

If an extreme type deviates to a platform that is more extreme than z_i^*, or between z_i^* and $\overline{z}_i$, this candidate is certain to lose because voters believe that such a candidate is an extreme type based on the off-path belief. Therefore, the expected utility is:

$$-p^M u\left(\left|\chi_j^M\left(z_j^*\right) - x_i^E\right|\right) - \sigma^M\left(1 - p^M\right)u\left(\left|\chi_j^E\left(z_j^*\right) - x_i^E\right|\right)$$

$$-\left(1 - \sigma^M\right)\left(1 - p^M\right)\int_{\overline{z}_j}^{\underline{z}_j} u\left(\left|\chi_j^E\left(z_j\right) - x_i^E\right|\right)dF\left(z_j\right)$$

Subtracting this equation from $V_i^E\left(z_i^*\right)$ yields

$$-u\left(\left|\chi_i^E\left(z_i^*\right) - x_i^E\right|\right) - c\left(\left|z_i^* - \chi_i^E\left(z_i^*\right)\right|\right) + \frac{p^M}{p^M + \sigma^M\left(1 - p^M\right)}$$

$$u\left(\left|\chi_j^M\left(z_j^*\right) - x_i^E\right|\right) + \frac{\sigma^M\left(1 - p^M\right)}{p^M + \sigma^M\left(1 - p^M\right)}u\left(\left|\chi_j^E\left(z_j^*\right) - x_i^E\right|\right)$$

A moderate type is indifferent to winning and losing at z_i^*. That is, (3.7) holds. Thus, from Lemma 3.4, the value of the above is positive, and this deviation decreases the expected utility. Note that (3.10) in the proof of Lemma 3.4 uses p^M, but the same result holds when p^M is replaced by $p^M / \left[p^M + \sigma^M \left(1 - p^M\right) \right]$.

A Moderate type does not deviate

Suppose R, without loss of generality. As a moderate type is indifferent between winning and losing at z_R^*, she is indifferent on whether to deviate to a platform that is more extreme than z_R^* or between z_R^* and $\bar{z}_R$. The second possibility involves deviating to any platform in $z_R' \in [\underline{z}_R, \bar{z}_R]$. For an extreme type, the candidate is indifferent between z_R^* and z_R'. This means that

$$
\begin{aligned}
&\left[p^M + \sigma^M \left(1 - p^M\right) \right]\left[u\left(x_R^E - \chi_R^E \left(z_R'\right)\right) + c\left(x_R^E \left(z_R'\right) - z_R'\right) \right] \\
&- \frac{1}{2}\left[\begin{array}{l} p^M u\left(x_R^E - \chi_L^M \left(z_L^*\right)\right) + \sigma^M \left(1 - p^M\right) u\left(x_R^E - \chi_L^E \left(z_L^*\right)\right) \\ + \left[p^M + \sigma^M \left(1 - p^M\right) \right]\left[u\left(x_R^E - \chi_R^E \left(z_R^*\right)\right) + c\left(\chi_R^E \left(z_R^*\right) - z_R^*\right) \right] \end{array} \right] \\
&= \left(1 - \sigma^M\right)\left(1 - p^M\right) \int_{\bar{z}_L}^{z_L} u\left(x_R^E - \chi_L^E \left(z_L\right)\right) dF\left(z_L\right) \\
&\quad - \left(1 - \sigma^M\right)\left(1 - p^M\right)\left(1 - F\left(z_L'\right)\right)\left[\begin{array}{l} u\left(x_R^E - \chi_R^E \left(z_R'\right)\right) \\ + c\left(\chi_R^E \left(z_R'\right) - z_R'\right) \end{array} \right]
\end{aligned}
\tag{3.18}
$$

A moderate type has no incentive to deviate to z_i' if

$$
\begin{aligned}
&\left[p^M + \sigma^M \left(1 - p^M\right) \right]\left[u\left(x_R^M - \chi_R^M \left(z_R'\right)\right) + c\left(\chi_R^M \left(z_R'\right) - z_R'\right) \right] \\
&- \frac{1}{2}\left[\begin{array}{l} p^M u\left(x_R^M - \chi_L^M \left(z_L^*\right)\right) + \sigma^M \left(1 - p^M\right) u\left(x_R^M - \chi_L^E \left(z_L^*\right)\right) \\ + \left[p^M + \sigma^M \left(1 - p^M\right) \right]\left[u\left(x_R^M - \chi_R^M \left(z_R^*\right)\right) + c\left(\chi_R^M \left(z_R^*\right) - z_R^*\right) \right] \end{array} \right] \\
&> \left(1 - \sigma^M\right)\left(1 - p^M\right) \int_{\bar{z}_L}^{z_L} u\left(x_R^M - \chi_L^E \left(z_L\right)\right) dF\left(z_L\right) \\
&\quad - \left(1 - \sigma^M\right)\left(1 - p^M\right)\left(1 - F\left(z_L'\right)\right)\left[\begin{array}{l} u\left(x_R^M - \chi_R^M \left(z_R'\right)\right) \\ + c\left(\chi_R^M \left(z_R'\right) - z_R'\right) \end{array} \right]
\end{aligned}
\tag{3.19}
$$

I disregard $\left(1 - \sigma^M\right)\left(1 - p^M\right)$ and differentiate the right-hand side of the above equations with respect to x_R^t to obtain $\int_{\bar{z}_L}^{z_L} u'\left(x_R^t - \chi_L^E \left(z_L\right)\right) dF\left(z_L\right) - \left(1 - F\left(z_L'\right)\right) u'\left(\left\| \chi_R^t \left(z_R'\right) - x_R^t \right\| \right)$. This is positive because the opponent's implemented policy is further from the ideal policy than z'_R, so the right-hand side of (3.18) is greater than the right-hand side of (3.19). From (3.15), at $z'_R = \bar{z}_R$, the left-hand side of (3.18) is zero. From (3.7), the left-hand side of (3.19) is

$\left[p^M + \sigma^M \left(1 - p^M \right) \right] \left[u \left(x^M_R - \chi^M_R \left(z'_R \right) \right) + c \left(\chi^M_R \left(z'_R \right) - z'_R \right) \right] + \sigma^M \left(1 - p^M \right) \left[u \left(x^M_R - \chi^M_R \left(z^*_R \right) \right) + c \left(\chi^M_R \left(z^*_R \right) - z^*_R \right) \right]$, so is positive because z'_R is

smaller than z^*_R. I differentiate the left-hand side with respect to z'_R. Note that σ^M and z^*_R are already decided, so only z'_R changes. Then,

$$\left[p^M + \sigma^M \left(1 - p^M \right) \right] \left[\begin{array}{c} -u' \left(\left| \chi^t_R \left(z'_R \right) - x^t_R \right| \right) \left(\dfrac{\partial \chi^t_R \left(z'_R \right)}{\partial z'_R} \right) + c' \left(\left| z'_R - \chi^t_R \left(z'_R \right) \right| \right) \\ \left(\dfrac{\partial \chi^t_R \left(z'_R \right)}{\partial z'_R} \right) - c \left(\left| z'_R - \chi^t_R \left(z'_R \right) \right| \right) \end{array} \right]$$

I ignore $p^M + \sigma^M \left(1 - p^M \right)$. From Lemma 3.1, this is negative. That is, $-u' \left(\left| \chi^t_R \left(z'_R \right) - x^t_R \right| \right) < 0$. This implies that if z'_R becomes smaller, then the left-hand sides of both equations increase. The next problem is the degree of increase. Differentiating $-u' \left(\left| \chi^t_R \left(z'_R \right) - x^t_R \right| \right)$ with respect to x^t_i yields

$$-u'' \left(\left| \chi^t_R \left(z'_R \right) - x^t_R \right| \right) \left(1 - \frac{\partial \chi^t_R \left(z'_R \right)}{\partial x^t_R} \right) \tag{3.20}$$

I differentiate (3.15) with respect to x^t_R, then

$$\frac{\partial \chi^t_R \left(z'_R \right)}{\partial z'_R} = \frac{u'' \left(\left| \chi^t_R \left(z'_R \right) - x^t_R \right| \right) c' \left(\left| z'_R - \chi^t_R \left(z'_R \right) \right| \right)}{\left[\begin{array}{c} u'' \left(\left| \chi^t_R \left(z'_R \right) - x^t_R \right| \right) c' \left(\left| z'_R - \chi^t_R \left(z'_R \right) \right| \right) \\ + u' \left(\left| \chi^t_R \left(z'_R \right) - x^t_R \right| \right) c'' \left(\left| z'_R - \chi^t_R \left(z'_R \right) \right| \right) \end{array} \right]} \in (0,1).$$

Thus, the value of (3.20) is negative. This implies that if x^t_R is more extreme, the increase of the left-hand side is lower when z'_R becomes smaller. At $z'_R = \bar{z}_R$, the left-hand side of (3.18) is lower than the right-hand side of (3.19). If z'_R becomes more moderate, both left-hand sides increase, but an increase in (3.19) is greater than an increase in (3.18). As a result, for all z'_R, the left-hand side of (3.18) is lower than the right-hand side of (3.19), and (3.19) is satisfied.

Finally, since a moderate type has no incentive to deviate to $\underline{z}_R$, she does not deviate to any policy that is more moderate than $\underline{z}_R$.

A two-policy semi-separating equilibrium

When (3.9) holds, a two-policy semi-separating equilibrium exists. When an extreme type chooses z_i^*, the expected utility is

$$V_i^E\left(z_i^*\right) = \left(\frac{1}{2}\right)\left[-p^M u\left(\left|\chi_j^M\left(z_j^*\right) - x_i^E\right|\right) - \sigma^M\left(1 - p^M\right) u\left(\left|\chi_j^E\left(z_j^*\right) - x_i^E\right|\right)\right.$$

$$\left. -\left\{p^M + \sigma^M\left(1 - p^M\right)\right\}\left\{u\left(\left|\chi_i^E\left(z_i^*\right) - x_i^E\right|\right) + c\left(\left|z_i^* - \chi_i^E\left(z_i^*\right)\right|\right)\right\}\right] - \left(1 - \sigma^M\right)$$

$u\left(\left|\chi_j^E\left(\bar{z}_j\right) - x_i^E\right|\right)$ The expected utility when the candidate chooses

$\bar{z}_i$ is $V_i^E\left(\bar{z}_i\right) = \left[p^M + \sigma^M\left(1 - p^M\right)\right]\left[u\left(\left|\chi_i^E\left(\bar{z}_i\right) - x_i^E\right|\right) - \left(c\left|\bar{z}_i - \chi_j^E\left(\bar{z}_i\right)\right|\right)\right]$

$$-\left(1/2\right)\left(1 - \sigma^M\right)\left(1 - p^M\right)\left[u\left(\left|\chi_j^E\left(\bar{z}_j\right) - x_i^E\right|\right) + u\left(\left|\chi_i^E\left(\bar{z}_i\right) - x_i^E\right|\right) + c\left(\left|\bar{z}_i - \right.\right.\right.$$

$\chi_i^E\left(\bar{z}_i\right)\left|\right)\right]$. When $\sigma^M = 1$, $V_i^E\left(\bar{z}_i\right)$ is greater than $V_i^E\left(z_i^*\right)$ because we assume that (3.5) does not hold. Assume $\bar{\sigma}^M$, which satisfies (3.15). If (3.9) holds, then $V_i^E\left(\bar{z}_i\right)$ is less than $V_i^E\left(z_i^*\right)$ at $\bar{\sigma}^M$. When σ^M increases continuously from $\bar{\sigma}^M$, $V_i^E\left(\bar{z}_i\right)$ increases and $V_i^E\left(z_i^*\right)$ decreases continuously, so there exists a σ^M under which $V_i^E\left(\bar{z}_i\right) = V_i^E\left(z_i^*\right)$, and such σ^M should be higher than $\bar{\sigma}^M$.

Platform $\bar{z}_i$ should be such that $\chi_i^E\left(\bar{z}_i\right)$ is between $\chi_i^M\left(z_i^*\right)$ and $\chi_i^E\left(z_i^*\right)$ if $p^M > 0$ and $\sigma^M > 0$, because in this region, there exists a policy that voters prefer to the expected implemented policy of a candidate with z_i^*. Thus, $\chi_i^E\left(\bar{z}_i\right)$ is more extreme than $\chi_i^M\left(z_i^*\right)$.

An extreme type does not deviate for the reason explained in the previous discussions. If an extreme type deviates to a platform that is more moderate than $\bar{z}_i$, the expected utility changes by $\left(1/2\right)\left(1 - \sigma^M\right)\left(1 - p^M\right)\left[u\left(\left|\chi_j^E\left(\bar{z}_j\right) - x_i^E\right|\right) - u\left(\left|\chi_i^E\left(\bar{z}_i\right) - x_i^E\right|\right) - c\left(\left|\bar{z}_i - \right.\right.\right.$
$\chi_i^E\left(\bar{z}_i\right)\left|\right)\right]$. This is negative because (3.9) holds.

A moderate type does not deviate to a more extreme policy than $\bar{z}_i$ for the reason explained in the previous subsection. A moderate type does not deviate to $\bar{z}_i$ if

$$\left[p^M + \sigma^M\left(1 - p^M\right)\right]\left[u\left(\left|\chi_i^M\left(\bar{z}_i\right) - x_i^M\right|\right) + c\left(\left|\bar{z}_i - \chi_i^M\left(\bar{z}_i\right)\right|\right)\right]$$

$$- u\left(\left|\chi_i^M\left(z_i^*\right) - x_i^M\right|\right) - c\left(\left|z_i^* - \chi_i^M\left(z_i^*\right)\right|\right)$$

$$- \left(1 - \sigma^M\right)\left(1 - p^M\right)\frac{1}{2}\begin{bmatrix}u\left(\left|\chi_j^E\left(\bar{z}_j\right) - x_i^M\right|\right) - u\left(\left|\chi_i^M\left(\bar{z}_i\right) - x_i^M\right|\right) \\ - c\left(\left|\bar{z}_i - \chi_i^M\left(\bar{z}_i\right)\right|\right)\end{bmatrix} > 0 \quad (3.21)$$

As $\overline{z}_i$ is more moderate than z_i^*, $u\left(\left|\chi_i^M\left(\overline{z}_i\right)-x_i^M\right|\right)+c\left(\left|\overline{z}_i-\chi_i^M\left(\overline{z}_i\right)\right|\right.$ $\left.-u\left(\left|\chi_i^M\left(z_i^*\right)-x_i^M\right|\right)-c\left(\left|z_i^*-\chi_i^M\left(z_i^*\right)\right|\right)\right)$ is positive. For an extreme type, $u\left(\left|\chi_j^E\left(\overline{z}_j\right)-x_i^E\right|\right)-u\left(\left|\chi_i^E\left(\overline{z}_i\right)-x_i^E\right|\right)-c\left(\left|\overline{z}_i-\chi_i^E\left(\overline{z}_i\right)\right|\right)$ is negative because (3.9) holds. From Lemma 3.4, its value for a moderate type is lower than for an extreme type, so $u\left(\left|\chi_j^E\left(\overline{z}_j\right)-x_i^M\right|\right)-u\left(\left|\chi_i^M\left(\overline{z}_i\right)\right.\right.$ $\left.-x_i^M\right|\left.\right)-c\left(\left|\overline{z}_i-\chi_i^M\left(\overline{z}_i\right)\right|\right)$ is also negative for a moderate type. As a result, (3.21) is satisfied.

Asymmetric equilibrium

There does not exist an asymmetric equilibrium in which candidates choose asymmetric platforms or different values of σ^M or $F(.)$. First, suppose that the support of $F(.)$ is asymmetric. Then, the probability of winning is constant in some regions of the support for at least one candidate, and it cannot be an equilibrium for the reason explained in Lemma 3.10. This means that σ^M should also be symmetric. Second, suppose that moderate types' platforms are asymmetric. This means that the probability of winning for a moderate type is also asymmetric. Suppose moderate type R announces a more extreme platform than moderate type L, and so loses to L. In this case, extreme type R has no incentive to imitate moderate type R. The values of σ^M should be symmetric, so it cannot be a semi-separating equilibrium.

■

3.A.5 Proposition 3.8

In such a separating equilibrium, the utility of candidate i of type t when he/she wins is $-u\left(\left|\chi_i^t\left(z_i^t\right)-x_i^t\right|\right)-c\left(\left|z_i^t-\chi_i^t\left(z_i^t\right)\right|\right)$. Then, the utility of candidate i of type t when a *same-type* opponent (type t) wins is $-u\left(\left|\chi_j^t\left(z_j^t\right)-x_i^t\right|\right)$. I denote $\hat{z}_i^t$ as the cut-off platform under which both of these utilities are the same for a type t candidate, and are symmetric $\left(\left|x_m-\hat{z}_i^t\right|=\left|x_m-\hat{z}_j^t\right|\right)$. Figure 3.4 shows the positions of $\hat{z}_R^t$ and $\chi_R^t\left(\hat{z}_R^t\right)$. Then, the following lemma holds.

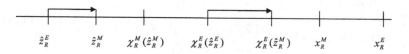

Figure 3.4 Separating Equilibrium.

An extreme type has an incentive to pretend to be moderate by choosing the moderate type's platform $\hat{z}_R^M$ because the probability of winning increases, and the implemented policy approaches the ideal policy.

Lemma 3.11

The extreme type's cut-off platform is more moderate than that of the moderate type, but the extreme type's implemented policy given the cut-off platform is more extreme than that of the moderate type.

Proof: To prove this, I differentiate (3.10) with respect to $x_R^t - x_L^t$ rather than x_R^t. Equation (3.10) can be rewritten as $u\left(x_R^t + \chi_R^t - 2x_m\right) - u\left(x_R^t - \chi_R^t\right) = c\left(\chi_R^t - z_R^t\left(\chi_R^t\right)\right)$, where $x_R^t - \chi_R^t = \left(x_R^t - x_m\right) + \left(\chi_R^t - x_m\right) = x_R^t + \chi_R^t - 2x_m$ because the platforms are symmetric. Differentiating both sides of (3.10) with respect to $x_R^t - x_L^t$ is the same as differentiating both sides of the rewritten equation with respect to x_R^t. Then,

$$\frac{\partial \chi_R^t}{\partial x_R^t} = \frac{\dfrac{u''\left(x_R^t - \chi_R^t\right)c'\left(\chi_R^t - z_R^t\left(\chi_R^t\right)\right)}{c''\left(\chi_R^t - z_R^t\left(\chi_R^t\right)\right)} - \left[u'\left(x_R^t - \chi_L^t\right) - u'\left(x_R^t - \chi_R^t\right)\right]}{u'\left(x_R^t - \chi_L^t\right) + u'\left(x_R^t - \chi_R^t\right)\dfrac{\partial z_R^t\left(\chi_R^t\right)}{\partial \chi_R^t}}.$$

For the same reason given in Lemma 3.4, this is positive. Note that $\partial z_R^t\left(\chi_R^t\right)/\partial x_R^t = -u''\left(x_R^t - \chi_R^t\right)/c''\left(\chi_R^t - z_R^t\left(\chi_R^t\right)\right) < 0$, and $\partial z_R^t\left(\chi_R^t\right) / \partial \chi_R^t = 1 + \left[u''\left(x_R^t - \chi_R^t\right)/c''\left(\chi_R^t - z_R^t\left(\chi_R^t\right)\right)\right] > 0$. Moreover,

$$\frac{\partial z_R^t\left(\chi_R^t\right)}{\partial x_R^t} + \frac{\partial z_R^t\left(\chi_R^t\right)}{\partial \chi_R^t}\frac{\partial \chi_R^t}{\partial x_R^t}$$

$$= -\frac{u''\left(x_R^t - \chi_R^t\right)}{c''\left(\chi_R^t - z_R^t\left(\chi_R^t\right)\right)}\left[1 - \frac{c'\left(x_R^t - z_R^t\left(\chi_R^t\right)\right)\dfrac{\partial z_R^t\left(\chi_R^t\right)}{\partial \chi_R^t}}{u'\left(x_R^t - \chi_L^t\right) + u'\left(x_R^t - \chi_R^t\right)\dfrac{\partial z_R^t\left(\chi_R^t\right)}{\partial \chi_R^t}}\right]$$

$$- \frac{u'\left(x_R^t - \chi_L^t\right) - u'\left(x_R^t - \chi_R^t\right)}{\dfrac{u'\left(x_R^t - \chi_L^t\right)}{\partial z_R^t\left(\chi_R^t\right)/\partial \chi_R^t} + u'\left(x_R^t - \chi_R^t\right)}$$

The second term of the right-hand side is negative. The first term is also negative since $u'\left(x_R^t - \chi_R^t\right) = c'\left(\chi_R^t - z_R^t\left(\chi_R^t\right)\right)$ from Lemma 3.1. Thus, this value is negative.

∎

Then, a symmetric separating equilibrium in which a moderate type wins does not exist. First, if a separating equilibrium in which a moderate type wins against an extreme type exists, regardless of off-path beliefs, an extreme type should announce $\hat{z}_i^E$. If the utility when an extreme candidate wins is higher than the utility when an extreme opponent wins, the extreme candidate has an incentive to win with certainty against the extreme opponent. This is made possible by approaching the median policy, regardless of off-path beliefs.

Second, a moderate type never announces a more moderate platform than $\hat{z}_i^E$. On such a platform, the utility when this moderate type candidate wins is lower than the utility when a moderate opponent wins. Therefore, the moderate type has an incentive to deviate and lose to the moderate opponent. If this moderate type also has an incentive to deviate and lose against an extreme opponent, she will deviate to lose with certainty. If this moderate type has an incentive to win against an extreme opponent, she will deviate to approach $\hat{z}_i^E$, because she can win against an extreme opponent and lose against a moderate opponent.

Finally, suppose that a moderate type announces a more extreme platform than $\hat{z}_i^E$. If an extreme type deviates to a moderate type's platform, the extreme type can improve his chance of winning and can implement a policy closer to his ideal. As a result, the extreme type can increase his expected utility from this deviation.

An asymmetric separating equilibrium in which a moderate type wins (z_R^M and z_L^M are asymmetric) also does not exist. Suppose that z_R^M and z_L^M are asymmetric, so one moderate candidate defeats a moderate opponent with certainty. Without loss of generality, suppose that moderate type R defeats moderate type L. That is, $\chi_R^M\left(z_R^M\right) - x_m < x_m - \chi_L^M\left(z_L^M\right)$, and moderate type R defeats extreme type L. Note that an extreme type announces $\hat{z}_i^E$. If z_R^M is more extreme than $\hat{z}_R^E\left(z_R^M > \hat{z}_R^E\right)$, an extreme type R will deviate to pretend to be a moderate type R. Therefore, assume that z_R^M is more moderate than $\hat{z}_R^E\left(z_R^M < \hat{z}_R^E\right)$. There are three cases.

The first case is that moderate type L loses to or has the same probability of winning as extreme type R $\left(\chi_R^E\left(\hat{z}_R^E\right) - x_m \leq x_m - \chi_L^M\left(z_L^M\right)\right)$. Regardless of off-path beliefs, if moderate type R's platform approaches $\hat{z}_R^E$, she can win against both moderate and extreme types of L, and the disutility following a win and the cost of betrayal decrease as the platform approaches her ideal policy.

The second case is that moderate type L defeats extreme type R $\left(\chi_R^E\left(\hat{z}_R^E\right) - x_m > x_m - \chi_L^M\left(z_L^M\right)\right)$ when moderate type L announces a more moderate platform than $\hat{z}_L^M$. From Lemma 3.4, moderate type R has an incentive to deviate and lose to moderate type L. If moderate type R approaches $\hat{z}_R^E$ by more than z_L^M, she can lose to moderate type L and still win against extreme type L.

The final case is that moderate type L defeats extreme type R $\left(\chi_R^E\left(\hat{z}_R^E\right)-x_m > x_m - \chi_L^M\left(z_L^M\right)\right)$ when moderate type L announces a platform that is the same as, or closer to her own ideal policy than $\hat{z}_L^M$. If extreme type L deviates to moderate type L's platform $\left(z_L^M\right)$, extreme type L can win against extreme type R with certainty and so gain a higher probability of winning. With this deviation, an extreme type can implement a policy closer to his ideal policy, so he will deviate.

∎

Notes

1 This chapter is revised version of Asako (2015).
2 This space is not a grid of evenly spaced policies like those introduced in Chapter 2. The policy space is continuous.
3 This section concentrates on a pooling equilibrium in which both types announce z_i^{M*}. The other possible pooling equilibria are discussed in Subsection 3.5.1.
4 To be precise, her equilibrium expected-payoff is
$$(1/2)\left[-p^M u\left(\left|\chi_j^M\left(z_j^{M*}\right)-x_i^M\right|\right)-(1-p^M)u\left(\left|\chi_j^E\left(z_j^{M*}\right)-x_i^M\right|\right)-u\left(\left|\chi_i^M\left(z_j^{M*}\right)-x_i^M\right|\right)-c\left(\left|z_i^{M*}-\chi_i^M\left(z_i^{M*}\right)\right|\right)\right]=-u\left(\left|\chi_i^M\left(z_j^{M*}\right)-x_i^M\right|\right)-c\left(\left|z_i^{M*}-x_i^M\left(z_j^{M*}\right)\right|\right)$$ since (3.2) holds. If a moderate type deviates by choosing any platform z_i which is more moderate than z_i^{M*}, and she wins with certainty, her expected payoff is $-u\left(\left|\chi_i^M\left(z_i\right)-x_i^M\right|\right)-c\left(\left|z_i-\chi_i^M\left(z_i\right)\right|\right)<-u\left(\left|\chi_i^M\left(z_i^{M*}\right)-x_i^M\right|\right)-c\left(\left|z_i^{M*}-\chi_i^M\left(z_i^{M*}\right)\right|\right)$.
5 There exist other semi-separating equilibria, as discussed in Subsection 3.5.1. The difference between the equilibrium in this subsection and the others is only that a moderate type chooses another platform. Thus, the basic characteristics are the same.
6 It is also possible to suppose that a policy space is discrete with a grid of policies similar to Chapter 2. That is, there are a large number of policy choices, and the distance between sequential policies is ϵ. If ϵ is very close to zero, thus approximating a continuous policy space, then there exists a most extreme platform that satisfies (3.8). The following results do not change given these settings.
7 Replace p^M by $p^M/\left(p^M+\sigma^M\left(1-p^M\right)\right)$ in (3.3). This result can then be derived in the same way as Lemma 3.4.

8 For example, suppose that off-path beliefs are $p_i(M|z_i) = p^M$ $\left(p^M \Big/ \left[p^M + \sigma^M\left(1-p^M\right)\right]\right)$ if the platform is more extreme than z_i^{M*} $\left(z_i^*\right)$, a moderate type has an incentive to compromise until $z_i^{M*}\left(z_i^*\right)$.

9 Criteria D1 and D2 in Cho and Kreps (1987) have the same result as universal divinity. On the other hand, divinity in Banks and Sobel (1987) cannot reduce the number of equilibria because it supports $p_i(M|z_i) = 0$ for any off-path platform.

10 In these examples, the decisions of "parties" are considered instead of "candidates." I use "candidates" in the model, but this can be replaced by "parties." My model supposes majority rule, so only the winner will be in the government. However, in national elections, the losing party will have some share of the sheets in Congress. If a party gets the majority of sheets, I take this to mean that the party wins over the opposition.

11 These problems are not only applied to my model but also to most voting models in the Downsian tradition.

12 To determine the mixed strategy in a continuous semi-separating equilibrium, I build on techniques introduced by Burdett and Judd (1983). They consider price competition and show that firms randomize prices when there is a possibility that consumers will observe only one price. Just as Burdett and Judd (1983) show that firms are indifferent over a range of prices, I show that an extreme type is indifferent over a range of platforms.

13 When extreme type R chooses $z'_R \in (\underline{z}_R, \overline{z}_R)$, the expected utility is

$$\left(1-p^M\right)F(z'_R)\left[-u\left(x_R^E - \chi_R^E(z'_R)\right) - c\left(\chi_R^E(z'_R) - z'_R\right)\right] - \left(1-\sigma^M\right)\left(1-p^M\right)\int_{z'_L}^{z_L} u\left(x_R^E - \chi_L^E(z_L)\right)dF(z_L).$$

Bibliography

Asako, Y., 2015, "Campaign Promises as an Imperfect Signal: How Does an Extreme Candidate Win against a Moderate One?," *Journal of Theoretical Politics* 27(4), pp. 613–649. (https://doi.org/10.1177/0951629814559724)

Banks, J., 1990, "A Model of Electoral Competition with Incomplete Information," *Journal of Economic Theory* 50(2), pp. 309–325. (https://doi.org/10.1016/0022-0531(90)90005-5)

Banks, J., and J. Sobel, 1987, "Equilibrium Selection in Signaling Games," *Econometrica* 55(3), pp. 647–661. (https://doi.org/10.2307/1913604)

Burdett, K., and K. Judd, 1983, "Equilibrium Price Dispersion," *Econometrica* 51(4), pp. 955–970. (https://doi.org/10.2307/1912045)

Callander, S., and S. Wilkie, 2007, "Lies, Damned Lies and Political Campaigns," *Games and Economic Behavior* 60(2), pp. 262–286. (https://doi.org/10.1016/j.geb.2006.12.003)

Cho, I., and D. Kreps, 1987, "Signaling Games and Stable Equilibria," *Quarterly Journal of Economics* 102(2), pp. 179–222. (https://doi.org/10.2307/1885060)

Clarke, P., 2004, *Hope and Glory: Britain 1900–2000*, London: Penguin Books.

Dağt, İ., 2006, "The Justice and Development Party: Identity, Politics, and Human Rights Discourse in the Search for Security and Legitimacy," in M. Yavuz ed., *The Emergence of a New Turkey: Democracy and the AK Party*, Salt Lake City: The University of Utah Press.

Huang, H., 2010, "Electoral Competition When Some Candidates Lie and Others Pander," *Journal of Theoretical Politics* 22(3), pp. 333–358. (https://doi.org/10.1177/0951629810365151)

Mulgan, A., 2002, *Japan's Failed Revolution: Koizumi and the Politics of Economic Reform*, Canberra: Asia Pacific Press.

4 Electoral promises with vague words

4.1 Introduction

Political parties and candidates usually prefer making ambiguous promises, a practice referred to as "political ambiguity." Prior studies usually interpret political ambiguity as a lottery: a probability distribution on single policies. In accordance with these past studies, this chapter extends the standard Downsian model with fully office-motivated candidates to allow a candidate to choose a lottery, rather than a single policy.[1] Then, this chapter identifies the conditions under which candidates choose ambiguous promises in equilibrium. Past studies indicate many possible reasons that political ambiguity occurs (Subsection 1.3.5), but this chapter analyzes convex voter preference as a possible explanation (Zeckhauser, 1969; Shepsle, 1972; Aragones and Postlewaite, 2002). To simplify the analysis, this chapter considers completely binding platforms, that is, candidates announce a lottery as a platform before an election, and the winner implements the policy according to the probability distribution of the announced platform after he/she wins the election.

This chapter shows that candidates choose ambiguous promises in equilibrium when (i) voters have convex utility functions, (ii) candidates are uncertain about voters' preferences, and (iii) the distribution of voters' preferred policies are polarized. Therefore, for political ambiguity to be considered as an equilibrium phenomenon with convex utility functions, voters must be polarized and voting must be probabilistic.

Through this chapter, I do not intend to say that the convexity of utility functions is the only reason political ambiguity emerges; many reasonable mechanisms have been suggested. However, while prior studies recognize the convexity of a voter's utility function as one reason for the emergence of ambiguity, none shows it as an equilibrium phenomenon without any restrictions on a candidate's strategy. Thus, one of the main contributions of this chapter is to explicitly show additional conditions (probabilistic voting and polarized voters) in which candidates choose a vague promise in equilibrium, given voters' convex utility functions while past studies did not establish the existence of equilibria in which candidates announce ambiguous promises with convex utility functions.

4.2 Deterministic voting

4.2.1 Discrete space

First, this subsection presents the implications of the deterministic model as a benchmark. Let me denote X as the set of policies and define $g(x, y)$ as the majority margin for $x, y \in X$; the number of voters who prefer x to y minus the number of voters who prefer y to x, where x and y are single policies. A policy x is the Condorcet winner when $g(x, y) \geq 0$ for all $y \in X$ (Black, 1948). Let us denote ΔX as the set of probability distributions over X. $g(p, q)$ is defined as the majority margin for lotteries $p, q \in \Delta X$. I call a Condorcet winner on ΔX a Condorcet winning lottery, which is defined as follows:[2]

Definition 4.1

A Condorcet winning lottery is a lottery p, such that $g(p, q) \geq 0$ for all $q \in \Delta X$.

Suppose three policies, $X = \{L, M, R\}$, where an element of ΔX is $(p_L, p_M, p_R) \in \Delta X$, and $p_x \geq 0$ is the probability that $x \in X$ occurs, where $p_L + p_M + p_R = 1$. In addition, suppose there is a population of voters from mass one, divided into three discrete groups, l, m, and r, and the proportion of voters in each group is less than 1/2; that is, no group constitutes a majority. Denote the set of groups as $G = \{l, m, r\}$, and its element as $g \in G$. Suppose that members of each group have the following preference relations:

$$
\begin{aligned}
l &: L \succ M \succ R \\
m &: M \succ L \succeq R \\
r &: R \succ M \succ L
\end{aligned}
\tag{4.1}
$$

These preference relations satisfy single-peakedness. Furthermore, the median point is M, which is the Condorcet winner in Black (1948).[3]

These preference relations of voters in group g can be represented by the Von Neumann-Morgenstern utility function $u_g : X \rightarrow \{0, v, 1\}$, with $v \in (0,1)$. The function assigns the value one to the most preferred alternative, v to the second-best alternative, and zero to the worst alternative. A voter has a concave utility function if $v > 1/2$, a linear utility function if $v = 1/2$, and a convex utility function if $v < 1/2$. Note that if a member of group m has $L \sim R$, the utilities of both L and R are $v = u_m(L) = u_m(R)$. I refer to the lottery $(p_L, p_M, p_R) = (0,1,0)$ simply as M. Assume that voters vote sincerely; they vote for the most preferred lottery among the alternatives. Assume also that voters choose to abstain when they are indifferent. If the two lotteries receive the same number of votes, the winner is chosen with an equal probability (50% each). Then, the following proposition is attained. Note that the result does not change even if voters choose a lottery with an equal probability when they are indifferent.[4]

Proposition 4.2

A Condorcet winning lottery is M when $v \geq 1/2$, and does not exist when $v < 1/2$.

   Proof: See Appendix 4.A.1.

   The rationale is as follows. Policy M cannot be the Condorcet winning lottery if voters have convex utility functions (i.e., $v < 1/2$). If M is chosen, the utilities of voters in groups l, m, and r are v, 1, and v, respectively. On the other hand, if lottery q_1 with $(p_L, p_M, p_R) = (1/2, 0, 1/2)$ is chosen, the utilities of voters in l, m, and r are $1/2$, $v/2$, and $1/2$, respectively. Thus, if $v < 1/2$, the voters in l and r prefer q_1 to M, and M is defeated by q_1 in a pairwise election. Moreover, q_1 cannot be a Condorcet winning lottery. If lottery q_2 with $(p_L, p_M, p_R) = (2/3, 1/3, 0)$ is chosen, the utilities of voters in l, m, and r are $(2+v)/3, (1+2v)/3$, and $v/3$, respectively. Thus, the voters in l and m prefer q_2 to q_1. However, q_2 is also defeated by q_3 with $(p_L, p_M, p_R) = (0, 2/3, 1/3)$. As in these cases, for any lottery, there is another that will receive the majority's support. The sum of the probabilities of choosing each policy is one. Thus, at least one group has a positive probability of its best policy being chosen. This probability can be divided between the remaining two groups' most preferred policies. This new lottery can then defeat the original lottery. On the other hand, when $v \geq 1/2$, M is not defeated by q_1 (or any other lotteries).

   Such a preference cycle usually occurs when a policy space has multiple dimensions. Supposing that candidates can choose a lottery instead of a single policy, the space of lotteries has two dimensions, since p_L and p_M should be identified (and p_R is determined by

$p_R = 1 - p_L - p_M$), even though the dimension of a single policy is one. However, if voters have concave utility functions, all voters prefer the least risk, that is, to make a certain choice. If they have linear utility functions, voters in l and r are indifferent, but voters in m still prefer M to q_1 because their utility is maximized when M is chosen. Consequently, if voters have concave or linear utility functions, they all (weakly) prefer a less ambiguous choice, in which case the space is considered one dimensional $(p_L = 1, p_M = 1,$ or $p_R = 1)$. Therefore, a Condorcet winning lottery exists if $v \geq 1/2$.

On the other hand, if voters have convex utility functions, conflicts of interest will arise among them: voters in group m prefer M to q_1 because their utility is maximized when M is definitively chosen. However, voters in groups l and r prefer q_1 to M because q_1 is riskier. Thus, both the position of a lottery and its degree of ambiguity matter, and this multi-dimensional space induces the non-existence of a Condorcet winning lottery.

4.2.2 Continuous space

When voters have concave or linear utility, the median policy is still the Condorcet winning lottery within a continuous policy space. Suppose that the policy space is continuous, $X \subset \mathbb{R}$. There is a continuum of voters, and each voter i has a single-peaked preference with his/her ideal point (the most preferred policy) $x_i \in X$. Voter i's preference can be represented by the function $-u_i(|\chi - x_i|)$ where χ is the implemented policy. Since a voter has a single-peaked preference, it satisfies $u_i'(d) > 0$ for $d > 0$. Voter's ideal points are distributed on X, and the distribution function is continuous and strictly increasing, meaning that there exists a unique median policy denoted by x_m.

In a lottery p_l, the implemented policies are distributed on $[\underline{x_l}, \overline{x_l}]$, and the probability distribution is $H_l(.)$. The upper bound of the lottery is $\overline{x_l}$ and the lower bound is $\underline{x_l}$, that is, $\underline{x_l} \leq \overline{x_l}$. The probability distribution does not need to be continuous and strictly increasing. The expected utility of voter i for a lottery p_l is

$$U_i(p_l) = \int_{\underline{x_l}}^{\overline{x_l}} -u_i(|\chi - x_i|) dH_l(\chi).$$

Denote also

$$E(x_l) = \int_{\underline{x_l}}^{\overline{x_l}} \chi \, dH_l(\chi)$$

as the expected value of the implemented policy of a lottery p_l. Then, the following proposition is obtained.

Proposition 4.3

Suppose $u_i'(d) > 0$ and $u_i''(d) \geq 0$ for $d > 0$. A Condorcet winning lottery is the median policy (a lottery which gives a probability one to x_m), and it is unique.

   Proof: See Appendix 4.A.2.

   In words, when voters are either risk averse or risk neutral, the Condorcet winning lottery is still the median policy. However, it may not be true when voters are risk loving.

4.3 Probabilistic voting

4.3.1 Settings

Return to the discrete settings with three policies in Section 4.2.1. A Condorcet winning lottery does not exist when voters have convex utility functions, as in the case of multiple policy dimensions. One method of finding an equilibrium when there are multiple policy dimensions is to introduce probabilistic voting, in which candidates are uncertain about voters' preferences.

   The voters' preference relations on policies and utilities are the same as those in (4.1). However, suppose that the members of group m have $L \sim R$, such that the utilities from L and R are both v.[5] A continuum of voters is distributed to each group according to a probability mass function $f : G \rightarrow [0, 1/2)$, with $f(m) = \gamma$ and $f(l) = f(r) = (1-\gamma)/2$, where $\gamma \in [0, 1/2)$; that is a symmetric distribution. The parameter γ represents the degree of centralization of the voter distribution. Two candidates 1 and 2 simultaneously determine the weight to allocate to each policy, $\Theta_i = \left(\theta_i^L, \theta_i^M, \theta_i^R \right) \in \Delta X$ before the election, where $i = 1$ or 2. The value of $\theta_i^x \in [0, 1]$ is the weight assigned to policy $x \in X$, where $\theta_i^L + \theta_i^M + \theta_i^R = 1$. Note that Θ_i is not a mixed strategy on X, because the policy is chosen after the election, while in a mixed strategy, a policy is chosen before the election. The model also supposes that voters believe that the probability that policy x will be implemented after an election is the same as the weight on x; therefore, candidates can affect voters' beliefs by allocating weights, as Callander and Wilson (2008) modeled.

   Candidate i obtains the share of votes given by

$$\Pi\left(\Theta_i, \Theta_{-i}\right) = \sum_{g \in G} f(g) \pi\left(\sum_{x \in X} \theta_i^x u_g(x) - \sum_{x \in X} \theta_{-i}^x u_g(x)\right),$$

where Θ_{-i} is the lottery chosen by i's opponent, and θ_{-i}^x is the weight on policy x in the opponent's promise. Suppose that an office-motivated candidate i maximizes $\Pi\left(\Theta_i, \Theta_{-i}\right)$.

The function $\pi : \mathbb{R} \to [0,1]$ is strictly increasing $(\pi'(t) > 0)$, satisfying $\pi(t) + \pi(-t) = 1$ (thus, $\pi(0) = 1/2$), and is strictly concave $(\pi''(t) < 0)$ for all $t \in [0, \infty)$. Since $\pi(t) + \pi(-t) = 1$, $\pi'(t) = \pi'(-t)$ for all $t \in [0, \infty)$. Here, $\sum_{x \in X} \theta_i^x u_g(x)$ is the expected utility of a voter in group g when candidate i wins the election. In addition, $\sum_{x \in X} \theta_i^x u_g(x) - \sum_{x \in X} \theta_{-i}^x u_g(x)$ is the difference in the expected utility of a voter in group g between the promise of candidate i and that of his/her opponent. If this is positive (negative), candidate i's lottery gives a higher (lower) expected utility than that of his/her opponent. In the deterministic model, $\pi(t) = 1$ when $t > 0$, and $\pi(t) = 0$ when $t < 0$. However, in the case of probabilistic voting, even if $t > 0$, $\pi(t) \in (1/2, 1]$. One interpretation of this is that voters make decisions based not only on candidates' policies but also on other factors, and therefore, their voting behavior is probabilistic.

4.3.2 *Equilibrium with convergence*

There exist multiple equilibria in this game. In order to clarify a situation where both candidates choose an ambiguous promise, I use the following corollary to show equilibria where both candidates choose the same lottery $\left(\Theta_i = \Theta_{-i}\right)$, that is, both candidates converge.

Corollary 4.4

i *If $v < 1/2$ and*

$$\gamma \le \frac{1 - 2v}{3 - 4v}$$

an equilibrium with $\Theta_i = \Theta_{-i}$ must satisfy $\theta_i^M = \theta_{-i}^M = 0$. (ii) Otherwise, an equilibrium with $\Theta_i = \Theta_{-i}$ must satisfy $\theta_i^M = \theta_{-i}^M = 1$.

Proof: See Appendix 4.A.3.

When voters have convex utility functions $(v < 1/2)$, and the proportion of median voters (γ) is sufficiently small, an ambiguous lottery such as $\Theta_i = \Theta_{-i} = (1/2, 0, 1/2)$ can be an equilibrium. Note that there exist many equilibria with $\Theta_i = \Theta_{-i}$ and $\theta_i^M = \theta_{-i}^M = 0$ such that

$\Theta_i = \Theta_{-i} = (1/3, 0, 2/3)$. Otherwise, both candidates converge to the median policy M.

When voters have concave or linear utility functions $(v \geq 1/2)$, there are no conflicts of interest regarding the degree of ambiguity because all voters (weakly) prefer the lower degree of ambiguity. Therefore, the candidates should converge on M with certainty in equilibrium. This situation is the same as that of the Downsian model, and the median voter becomes critical in deciding the winner.

On the other hand, when voters have convex utility functions, conflicts of interest among the voters on the degree of ambiguity do exist, because the voters in groups l and r (extreme voters) prefer a higher degree of ambiguity, whereas those in group m (median voters) still prefer a less ambiguous policy. If the proportion of median voters γ is sufficiently high $(\gamma > (1-2v)/(3-4v))$, candidates need to consider the median voters' interests, and thus, they converge on the median policy. However, if γ is low $(\gamma < (1-2v)/(3-4v))$, candidates care more about the extreme voters than they do about median voters, and thus, choose an ambiguous policy.

In many extensions of the Downsian model of electoral competitions, the candidate who wins the support of the median voter is the winner. However, when (i) voters have convex utility functions and (ii) the proportion of median voters are small, a candidate cannot win even if he/she gets the support of the median voter. Rather, candidates must ignore the interests of the median voter to win the election.

4.3.3 Equilibrium with divergence

There also exist other equilibria with divergence, that is, $\Theta_i \neq \Theta_{-i}$. Denote

$$\bar{\theta} \equiv \theta_i^L - \theta_{-i}^L,$$

$$\bar{\gamma} \equiv \frac{(1-2v)\pi'(\bar{\sigma})}{2(1-v)\pi'(0) + (1-2v)\pi'(\bar{\sigma})}.$$

I then have the following proposition.[6]

Proposition 4.5

Suppose $v < 1/2$. A strategy profile with $\theta_i^M = \theta_{-i}^M = 0$ with $\theta_i^L - \theta_{-i}^L = \bar{\theta}$ (hence $\theta_i^R - \theta_{-i}^R = -\bar{\theta}$) for all i is a Nash equilibrium when $\gamma \leq \bar{\gamma}$.
    Proof: See Appendix 4.A.3.

As in Corollary 4.4, when the degree of political centralization γ is sufficiently small, political ambiguity can emerge. Note that when voters have concave or linear utility functions, both candidates choosing M definitively is a unique equilibrium.

Corollary 4.6

If $v \geq 1/2$, $\theta_1^M = \theta_2^M = 1$ is a unique equilibrium.
Proof: See Appendix 4.A.3.

The probabilistic voting model adopted here is based on that of Kamada and Kojima (2014), who suppose that candidates can choose only a single policy (not a lottery). They show that with convex utility functions of voters and a polarized voter distribution, perfectly divergent candidates result in a unique equilibrium. Here, perfect divergence means that, without exception, the left candidate chooses a left policy, while the right candidate chooses a right policy, that is, $\Theta_i = (1,0,0)$ and $\Theta_{-i} = (0,0,1)$. On the other hand, the model of this chapter allows candidates to choose a lottery instead of a single policy, which increases the number of equilibria. Thus, ambiguity can arise in the form of equilibrium strategies in the context of convex voter utilities. In some equilibria, candidates choose the same ambiguous lottery, so policy divergence does not occur. On the other hand, perfectly divergent equilibrium shown by Kamada and Kojima (2014) also exists in this model. Therefore, this model demonstrates that a probability voting model with convex utilities is useful to show not only political polarization but also political ambiguity, and candidates may choose partially divergent policies: they combine policy divergence and political ambiguity (i.e., $\theta_i^L > 0$ and $\theta_i^R > 0$ with $\Theta_i \neq \Theta_{-i}$).

In addition, the following corollary is obtained.

Corollary 4.7

As $\pi'(\bar{\theta})$ increases, $\bar{\gamma}$ increases. Therefore, (i) it is less likely to lead to an equilibrium with more divergence, that is, higher $\bar{\theta}$ and (ii) if voters are more sensitive to differences between candidates, candidates tend to be converged to the same promise (i.e., $\Theta_i = \Theta_{-i}$).

Because $\pi''(t) < 0$ for all $t \in [0,\infty)$, a policy with more divergence has lower $\pi'(\bar{\theta})$, which decreases $\bar{\gamma}$. Thus, the condition $\gamma \leq \bar{\gamma}$ becomes more difficult to satisfy, so the first implication (i) is obtained. To understand the second implication (ii), suppose two functions π and $\bar{\pi}$ such that $\pi(t) < \bar{\pi}(t)$ for all $t \in [0,\infty)$, that is, voters are more sensitive to policy divergence with $\bar{\pi}$ than π. Since $\pi(t) < \bar{\pi}(t)$ for all $t \in [0, \infty)$, $\pi'(t) < \bar{\pi}'(t)$ with low t while $\pi'(t) > \bar{\pi}'(t)$ with high t where $t \in [0, \infty)$. This means that the condition $\gamma \leq \bar{\gamma}$ is more likely to be satisfied with low $\bar{\theta}$ but becomes difficult to be satisfied with high $\bar{\theta}$.

According to Proposition 4.5, both policy divergence and political ambiguity can occur when voters have convex preferences with low γ. However, Corollary 4.6 indicates that political ambiguity is more likely to occur compared to policy divergence.

4.4 Application: the constitutional reform in Japan

When (i) voters have a convex utility function and (ii) the distribution of their most preferred policies is polarized, candidates choose policy divergence, political ambiguity, or any combination of the two. As discussed in Chapter 1, voters may have convex utility functions on non-economic issues. Although policy divergence is observed for some non-economic issues with polarized voters, such as the debate around same-sex marriage in the United States (Kamada and Kojima, 2014), candidates also prefer choosing an ambiguous position. Another example of political ambiguity is the constitutional reform in Japan.

The Constitution of Japan was enacted in 1947 as the new constitution for post-war Japan. In 1947, Japan was occupied by the Allies, mainly the United States. Thus, the Constitution was written by non-Japanese, although the opinions of many Japanese were considered. Therefore, constitutional reform has been a topic of frequent discussion since Japan gained independence. Article 9 is the most controversial, as it prohibits Japan from maintaining a military, air force, navy, or settling any international dispute using force. Nevertheless, Japan has had a defense force that has held military power since 1954. Public opinion on constitutional reform is divided. According to a 2017 poll conducted by NHK (the public broadcaster in Japan), 43% of the responses were in favor of the reform, while 34% of the responses were against the reform. These findings have been consistent over time.[7] This issue is not related to the economy; therefore, it may be a convex issue, and the distribution of voters' opinions is polarized. Thus, the conditions for political ambiguity are satisfied.

Since 1955, the Liberal Democratic Party of Japan (LDP) has run the government, except during the periods, 1993–1994 and 2008–2012. In the early period of the LDP administration (e.g., the Hatoyama administration, 1954–1956), many LDP members claimed that the Constitution should be written by the Japanese people. However, since the 1960s, LDP administrations have avoided discussing (and almost given up on) this issue because public opinion was so divided and an intra-LDP faction hesitated to implement reforms (Machidori, 2016, p. 4). Consequently, the Japanese Constitution has not yet been revised.

Recently, Prime Minister Shinzo Abe explicitly promised to reform the Constitution in the 2017 general election. The 2017 LDP manifesto (a booklet containing campaign promises) devoted two pages (out of a total of 38 pages) to this promise. In contrast, in the 2012 and 2014 elections, in which Abe was also the party leader, the LDP manifestos devoted only one-sixth to half a page (out of 26 pages) to this issue. Moreover, even in the 2017 manifesto, details on the reform remained vague since most parts only show some images of the constitutional reform. Voters believed that the LDP was more likely to revise the Constitution than its opponents $\left(\theta_i^R > \theta_i^L \right)$; however, it was ambiguous $\left(\theta_i^R \neq 1 \right)$.

In an election, parties and candidates usually announce and promote their economic policies. Indeed, most LDP manifestos in 2012, 2014, and 2017 laid significant emphasis (large weights) on explaining economic policies (popularly known as Abenomics in Japan). On the other hand, candidates prefer maintaining a degree of ambiguity on social and national-security issues. Possibly, they prefer specifying an economic policy because voters have concave utility functions for such economic policies. However, they prefer ambiguity for non-economic and polarized issues, where voters may have convex utility functions.

4.5 Summary

Prior studies usually interpret political ambiguity as a lottery. This chapter also supposes that voters can choose between lotteries, rather than a single policy. Further, it identifies the conditions under which political ambiguity occurs in equilibrium, given the convex utility functions of voters. In the deterministic model, if voters have concave or linear utility functions, the median policy is still the Condorcet winner. However, if voters have convex utility functions, the existence of the Condorcet winning lottery is not ensured because the space of campaign promises has multiple dimensions. On the other hand, in the probabilistic voting model, candidates choose an ambiguous promise in equilibrium when (i) voters have convex utility functions and (ii) the distribution of voters' most preferred policies is polarized. Therefore, to have political ambiguity as an equilibrium phenomenon with convex utility functions of voters, voters need to be polarized, and candidates must be uncertain about voters' preferences.

4.A Appendix: proofs

4.A.1 Proposition 4.2

First, suppose a lottery q_L with ($p_L > 0, p_M \geq 0,\ p_R \geq 0$), and a second lottery q'_L with ($p'_L = 0, p'_M \geq 0,\ p'_R \geq 0$). When a member of group m has $L \succ R$, the voters in m prefer q'_L to q_L if $p_L v + (1 - p_L - p_R) < 1 - p'_R$, which is $p_R + p_L(1 - v) > p'_R$. When a member of group m has $L \sim R$, they prefer q'_L to q_L if $p_L v + (1 - p_L - p_R) + p_R v < (1 - p'_R) + p'_R v$, which is $p_R + p_L > p'_R$. On the other hand, the voters in r prefer q_L to q_L if $(1 - p_L - p_R)v + p_R < (1 - p'_R)v + p'_R$, which is $p'_R > p_R - (p_L v)/(1 - v)$. Because $v \in (0,1)$, $p_L + p_R > p_L(1 - v) + p_R > p_R - (p_L v)/(1 - v)$, which means there exists q'_L, which defeats q_L. Thus, no lottery with $p_L > 0$ can be a Condorcet winning lottery.

Second, suppose a lottery q_R with ($p_L \geq 0, p_M \geq 0,\ p_R > 0$), and a second lottery q'_R with ($p'_L \geq 0, p'_M \geq 0, p'_R = 0$). When a member of group m has $L \succ R$, the voters in m prefer q'_R to q_R if $p_L v + (1 - p_L - p_R) < p'_L v + (1 - p'_L)$, which is $p_L + p_R/(1 - v) > p'_L$. When a member of m has $L \sim R$, they prefer q'_R to q_R if $p_L v + (1 - p_L - p_R) + p_R v < p'_L v + (1 - p'_L)$, which is $p_L + p_R > p'_L$. On the other hand, the voters in l prefer q'_R to q_R if $p_L + (1 - p_L - p_R)v < p'_L + (1 - p'_L)v$, which is $p_L - (p_R v)/(1 - v) < p'_L$. Because $v \in (0,1)$, $p_L + p_R > p_L + p_R/(1 - v) > p_L - (p_R v)/(1 - v)$, there exists q'_R, which defeats q_R. Thus, no lottery with $p_R > 0$ can be a Condorcet winning lottery.

Therefore, only $M\left((p_L, p_M, p_R) = (0,1,0)\right)$ can be a Condorcet winning lottery. From M, the utilities of the voters in l, m, and r are v, 1, and v, respectively. Because members of m earn the highest utility from M, they do not have an incentive to deviate. Suppose I have another lottery q_M with (p'_L, p'_M, p'_R). Then, the utilities of the voters in l and r are $p'_L + v(1 - p'_L - p'_R)$ and $p'_R + v(1 - p'_L - p'_R)$, respectively, from this lottery. If $p'_L \leq v$ or $p'_R \leq v$, q_M would never be able to defeat M. On the other hand, if $p'_L > v$ and $p'_R > v$, q_M can defeat M because the members of l and r prefer q_M to M. The conditions $p'_L > v$ and $p'_R > v$ can be satisfied at the same time only if $v < 1/2$. Therefore, a Condorcet winning lottery is M if and only if $v \geq 1/2$, and does not exist otherwise.

∎

4.A.2 Proposition 4.3

Consider that there are two lotteries, p_1 which gives probability 1 to x_m and p_2. First, suppose that p_2 also gives probability 1 to x_m. Then, all voters are indifferent between p_1 and p_2, and thus they tie.

Second, suppose p_2 with $\underline{x}_2 < \bar{x}_2$ and $E(x_2) = x_m$. If $u_i''(d) \geq 0$ for $d > 0$, the following equation holds:

$$-\int_{\underline{x}_2}^{\bar{x}_2} u_i\left(|\chi - x_i|\right) dH_2\left(\chi\right) \leq -u_i\left(\int_{\underline{x}_2}^{\bar{x}_2} |\chi - x_i| dH_2\left(\chi\right)\right)$$
$$\leq -u_i\left(|E(x_2) - x_i|\right) \qquad (4.2)$$

The second inequality is from the characteristic of an absolute value, that is, $\left|E(y)\right| < E(|y|)$ since y may take a negative value. When voter i has $u_i''(d) > 0$, the first inequality holds as a strict inequality from the Jensen's inequality. Since $u_i\left(|E(x_2) - x_i|\right) = u_i\left(|x_m - x_i|\right)$, such voters strictly prefer p_1 to p_2. If voter i has $u_i''(d) = 0$, the first inequality of (4.2) becomes equality. However, the second inequality still holds. Since this voter has a linear preference, voter i whose ideal point is not within $\left[\underline{x}_2, \bar{x}_2\right]$ ($x_i \notin \left[\underline{x}_2, \bar{x}_2\right]$) is indifferent to both lotteries, so they abstain. For voters whose ideal point is within $\left[\underline{x}_2, \bar{x}_2\right]$ ($x_i \in \left[\underline{x}_2, \bar{x}_2\right]$), the second inequality of (4.2) holds as strict inequality. Thus, voters whose ideal point is within $\left[\underline{x}_2, \bar{x}_2\right]$ strictly prefer p_1 to p_2. As a result, when all voters have $u_i''(d) \geq 0$, p_1 wins against p_2.

The final case is p_2 with $\underline{x}_2 < \bar{x}_2$ and $E(x_2) \neq x_m$. Suppose $E(x_2) < x_m$. From the above analyses, voters whose ideal point is higher than x_m ($x_i \geq x_m$) strictly prefers p_1 to p_2. Moreover, voters whose ideal point is $x_i \in \left(E(x_2), x_m\right)$ are divided into two groups, namely, those that support p_1 and those that support p_2. Thus, the majority of voters strictly prefer p_1 to p_2. When $E(x_2) > x_m$, the result is the same.

Since any lottery except x_m is defeated by x_m, and x_m ties with x_m, the median policy is the unique Condorcet winning lottery.

∎

4.A.3 Proposition 4.5, and Corollaries 4.4 and 4.6

First, I obtain the following lemma. Denote $\alpha \in [0,1]$, the weight on policy L such that $\theta_i^L = \alpha\left(1 - \theta_i^M\right)$ and $\theta_i^R = (1-\alpha)\left(1 - \theta_i^M\right)$.

Lemma 4.A

Suppose $\theta_i^M = \theta_{-i}^M$. Then, given the opponent's strategy, and θ_i^M, a candidate is indifferent among any $\Theta_i = \left(\theta_i^L, \theta_i^M, \theta_i^R\right) = \left(\alpha\left(1 - \theta_i^M\right), \theta_i^M, (1-\alpha)\left(1 - \theta_i^M\right)\right)$ with $\alpha \in [0,1]$.

Proof: Suppose that the opponent of candidate i chooses Θ_{-i}, and from this policy, the voters in each group get u_l, u_m, and u_r, respectively. Then, candidate i chooses $\Theta_i = \left(\theta_i^L, \theta_i^M, \theta_i^R\right)$, such that it maximizes

$$
\begin{aligned}
\Pi(\Theta_i, \Theta_{-i}) = & \frac{1-\gamma}{2}\pi\left(\alpha\left(1-\theta_i^M\right)+\theta_i^M v - u_l\right) + \\
& \gamma\pi\left(\theta_i^M + \left(1-\theta_i^M\right)v - u_m\right) \\
& + \frac{1-\gamma}{2}\pi\left((1-\alpha)\left(1-\theta_i^M\right)+\theta_i^M v - u_r\right)
\end{aligned}
\tag{4.3}
$$

where $\alpha \in [0,1]$. Suppose $\theta_i^M = \theta_{-i}^M$, and denote $\hat{\theta} \equiv \alpha\left(1-\theta_i^M\right)+\theta_i^M v - u_l$. Then, $(1-\alpha)\left(1-\theta_i^M\right)+\theta_i^M v - u_r = -\hat{\theta}$ when $\left(1-\theta_i^M\right)+2\theta_i^M v = u_l + u_r$ is satisfied. Because $u_l = \theta_{-i}^L + \theta_{-i}^M v$ and $u_r = \theta_{-i}^R + \theta_{-i}^M v$, $u_l + u_r = \left(1-\theta_{-i}^M\right)+2\theta_{-i}^M v$. Therefore, when $\theta_i^M = \theta_{-i}^M$, $\left(1-\theta_i^M\right)+2\theta_i^M v = u_l + u_r$ and $\theta_i^M + \left(1-\theta_i^M\right)v - u_m = 0$ are satisfied. Then, (4.3) can be written to

$$
\Pi(\Theta_i, \Theta_{-i}) = \frac{1-\gamma}{2}\pi\left(\hat{\theta}\right)+\gamma\pi(0)+\frac{1-\gamma}{2}\pi\left(-\hat{\theta}\right)
$$

Because $\pi(t)+\pi(-t)=1$ for any $t \in [0,\infty)$, it becomes

$$
\Pi(\Theta_i, \Theta_{-i}) = \frac{1-\gamma}{2}+\gamma\pi(0).
$$

which does not depend on α. It means that the probability of winning does not depend on a value of α when $\theta_i^M = \theta_{-i}^M$.

■

Differentiate (4.3) with respect to θ_i^M. Then, the first derivative is

$$
\begin{aligned}
\frac{\partial \Pi(\Theta_i, \Theta_{-i})}{\partial \theta_i^M} = & -\frac{1-\gamma}{2}(\alpha - v)\,\pi'\left(\alpha\left(1-\theta_i^M\right)+\theta_i^M v - u_l\right) \\
& + \gamma(1-v)\,\pi'\left(\theta_i^M + \left(1-\theta_i^M\right)v - u_m\right) \\
& - \frac{1-\gamma}{2}\left[(1-\alpha)-v\right]\pi'\left((1-\alpha)\left(1-\theta_i^M\right)+\theta_i^M v - u_r\right).
\end{aligned}
\tag{4.4}
$$

Suppose $\theta_i^M = \theta_{-i}^M$. From the proof of Lemma 4.A, $\alpha\left(1-\theta_i^M\right)+\theta_i^M v - u_l = -\left[(1-\alpha)\left(1-\theta_i^M\right)+\theta_i^M v - u_r\right]$, so

$$
\pi'\left(\alpha\left(1-\theta_i^M\right)+\theta_i^M v - u_l\right) = \pi'\left((1-\alpha)\left(1-\theta_i^M\right)+\theta_i^M v - u_r\right)
$$

since $\pi'(t) = \pi'(-t)$ for any $t \in [0, \infty)$. Thus, the first derivative becomes

$$
\begin{aligned}
\frac{\partial \Pi(\Theta_i, \Theta_{-i})}{\partial \theta_i^M} &= -\frac{1-\gamma}{2}(1-2v)\,\pi'\left(\alpha\left(1-\theta_i^M\right)+\theta_i^M v - u_l\right) \\
&\quad + \gamma(1-v)\,\pi'\left(\theta_i^M + \left(1-\theta_i^M\right)v - u_m\right)
\end{aligned}
\tag{4.5}
$$

Suppose $v < 1/2$. A candidate will choose $\theta_i^M = \theta_{-i}^M = 0$ (when (4.5) is strictly negative) or $\theta_i^M = \theta_{-i}^M = 1$ (when (4.5) is strictly positive) when they are not indifferent. Suppose $\theta_i^M = \theta_{-i}^M = 0$. Then, from $\bar{\theta} \equiv \theta_i^L - \theta_{-i}^L$, (4.5) becomes

$$
\frac{\partial \Pi(\Theta_i, \Theta_{-i})}{\partial \theta_i^M} = -\frac{1-\gamma}{2}(1-2v)\,\pi'\left(\bar{\theta}\right) + \gamma(1-v)\,\pi'(0)
$$

Then, $\partial \Pi(\Theta_i, \Theta_{-i}) / \partial \theta_i^M \leq 0$ when

$$
\gamma \leq \frac{(1-2v)\,\pi'\left(\bar{\theta}\right)}{2(1-v)\,\pi'(0) + (1-2v)\,\pi'\left(\bar{\theta}\right)} \equiv \bar{\gamma}
\tag{4.6}
$$

is satisfied. Because $\partial \Pi(\Theta_i, \Theta_{-i}) / \partial \theta_i^M \leq 0$, a lower θ_i^M gives a (weakly) higher vote share. Therefore, $\theta_i^M = 0$ is the best response for candidate i. Thus, $\theta_i^M = \theta_{-i}^M = 0$ with $\theta_i^L - \theta_{-i}^L = \bar{\theta}$ is an equilibrium when $v < 1/2$ and $\gamma \leq \bar{\gamma}$ (Proposition 4.5).

In an equilibrium with convergence $(\Theta_i = \Theta_{-i})$, $\bar{\theta} = 0$ is satisfied. Then, (4.6) can be rewritten as

$$
\gamma \leq \frac{(1-2v)\,\pi'(0)}{2(1-v)\,\pi'(0) + (1-2v)\,\pi'(0)} = \frac{1-2v}{3-4v}
$$

Thus, if $\gamma \leq (1-2v)/(3-4v)$, a lower θ_i^M gives a (weakly) higher vote share, so $\theta_i^M = \theta_{-i}^M = 0$ in equilibrium (Corollary 4.4(i)).

If $v \geq 1/2$, (4.5) is strictly positive for any value of γ. Thus, regardless of the opponent's strategy, $\Theta_i = (0,1,0)$ is the best response.

Note that candidates may choose $\theta_i^M \neq \theta_{-i}^M$. A candidate chooses $\theta_i^M \in (0,1)$ when a candidate is indifferent, that is, (4.4) is exactly the same as zero, but I ignore such an indifferent case since it rarely occurs with a specific value of γ. The remaining case of $\theta_i^M \neq \theta_{-i}^M$ is $\theta_i^M = 0$ and $\theta_{-i}^M = 1$. The vote share (4.3) becomes

$$
\Pi(\Theta_i, \Theta_{-i}) = \frac{1-\gamma}{2}\pi(\alpha - v) + \gamma\pi(v-1) + \frac{1-\gamma}{2}\pi(1-\alpha - v)
$$

Differentiate (4.3) with respect to α. Then, the first-order condition is

$$\frac{\partial\Pi(\Theta_i,\Theta_{-i})}{\partial\alpha} = \frac{1-\gamma}{2}\pi'(\alpha-v) - \frac{1-\gamma}{2}\pi'(1-\alpha-v) = 0 \qquad (4.7)$$

and the second derivative is strictly negative since $\pi''(t) < 0$ for all $t \in [0,\infty)$. The opponent also has the same first-order condition. Condition (4.7) is satisfied if and only if $\alpha = 1/2$. Therefore, given $\theta_i^M = 0$ and $\theta_{-i}^M = 1$, both candidates choose $\alpha = 1/2$. When $\alpha = 1/2$, (4.4) becomes

$$\frac{\partial\Pi(\Theta_i,\Theta_{-i})}{\partial\theta_i^M} = -(1-\gamma)\left(\frac{1}{2}-v\right)\pi'\left(\frac{1}{2}-v\right) + \gamma(1-v)\pi'(v-1) \qquad (4.8)$$

for both candidates. In order to satisfy $\theta_i^M = 0$ and $\theta_{-i}^M = 1$, (4.8) must be strictly positive for one candidate and strictly negative for another candidate, but it is impossible. Thus, $\theta_i^M = 0$ and $\theta_{-i}^M = 1$ is not equilibrium regardless of the value of v. (Corollary 4.4(ii) and Corollary 4.6).

∎

Notes

1 This chapter is revised version of Asako (2019).
2 This differs from the maximal lottery (probabilistic/randomized Condorcet winner) proposed by Fishburn (1984). A maximal lottery supposes that voters make a decision after a policy is revealed from each lottery, whereas a Condorcet winning lottery supposes that voters choose before the outcomes of the lotteries are revealed. More precisely, p is a maximal lottery if $\sum_{x,y \in X} p(x)q(y)g(x,y) \geq 0$ for all $q \in \Delta X$.
3 If a Condorcet winner does not exist, a Condorcet winning lottery does not exist either (Fishburn, 1972).
4 In the following proposition, Shepsle (1972) demonstrates the case with risk-loving voters, and Aragones and Postlewaite (2002) demonstrate the case with risk-averse voters. I merge these two findings by adding an analysis with risk-neutral voters.
5 This assumption is not critical. Even if $L \succ R$ or $L \prec R$, the main implications do not change.
6 It does not consider an indifferent case which rarely occurs. See the appendix for more details.
7 From the website of NHK (https://www3.nhk.or.jp/news/special/kenpou70/yoron2017.html).

Bibliography

Aragones, E., and A. Postlewaite, 2002, "Ambiguity in Election Games," *Review of Economic Design* 7, pp. 233–255. (https://doi.org/10.1007/s100580200081)

Asako, Y., 2019, "Strategic Ambiguity with Probabilistic Voting," *Journal of Theoretical Politics* 31(4), pp. 626–641. (https://doi.org/10.1177/0951629819875516)

Black, D., 1948, "On the Rationale of Group Decision Making," *Journal of Political Economy* 56(1), pp. 23–34. (https://doi.org/10.1086/256633)

Callander, S., and C. Wilson, 2008, "Context-Dependent Voting and Political Ambiguity," *Journal of Public Economics* 92(3–4), pp. 565–581. (https://doi.org/10.1016/j.jpubeco.2007.09.002)

Fishburn, P. C., 1972, "Lotteries and Social Choice," *Journal of Economic Theory* 5(2), pp. 189–207. (https://doi.org/10.1016/0022-0531(72)90101-9)

Fishburn, P. C., 1984, "Probabilistic Social Choice Based on Simple Voting Comparison," *Review of Economic Studies* 51, pp. 683–692. (https://doi.org/10.2307/2297786)

Kamada, Y., and F. Kojima, 2014, "Voter Preferences, Polarization, and Electoral Policies," *American Economic Journal: Microeconomics* 6(4), pp. 203–236. (https://doi.org/10.1257/mic.6.4.203)

Machidori, S., 2016, "Seizigaku kara mita Kenpou Kaisei [The Constitutional Reform from the Point of View of Political Science]," in K. Komamura and S. Machidori eds., *Kenpou Kaisei no Hikaku Seizigaku* [*Comparative Politics of the Constitutional Reform*], Tokyo: Kobundo, pp. 2–18.

Shepsle, K., 1972, "The Strategy of Ambiguity: Uncertainty and Electoral Competition," *American Political Science Review* 66, pp. 555–568. (https://doi.org/10.2307/1957799)

Zeckhauser, R., 1969, "Majority Rule with Lotteries on Alternatives," *The Quarterly Journal of Economics* 83(4), pp. 696–703. (https://doi.org/10.2307/1885458)

Index